A Beginner's Handbook in Biological Electron Microscopy

A Beginner's Handbook in Biological Electron Microscopy

BRENDA S. WEAKLEY
A.B., A.M., Ph.D.

Lecturer in Anatomy, The University, Dundee

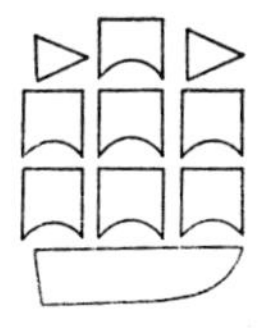

CHURCHILL LIVINGSTONE
EDINBURGH AND LONDON 1972

ISBN 0 443 00908 2

Printed in Great Britain by
Northumberland Press Limited, Gateshead

TO TIMOTHY

Preface

In the course, over the past few years, of teaching the theory and practical techniques of electron microscopy to candidates for the degree of Bachelor of Medical Science, it has become increasingly evident that there is no short, concise, and easily comprehended book on the market which deals with certain aspects which bother the beginner. No one, it seems, has produced a short book dealing adequately with the considerable no-man's-land which lies between theory and cookery. The necessary information exists in written form, but acquiring it demands weeks of reading from texts 400 to 500 pages in length together with the study of review articles and basic papers, the references to which are difficult for the beginner to find.

Many beginners, I believe, find the number of thick volumes which are available a bit daunting, as my own students and I have done. If one is to become a career electron microscopist one *must* read these books—but not at once. They will be better understood when the beginner has actually done some electron microscopy—processed some tissues, breathed a bit of osmium vapour inadvertently, and lost his fear of manipulating the electron microscope.

It was not until I started writing this book that I realized how superficial was my own knowledge of electron microscopy. Double checking my 'facts' revealed that many of them were myths. Seeking expert advice from the literature, I found that the experts disagree with alarming frequency. The end result is a book which, I hope, will give some helpful practical advice to the beginner who needs it, without leading him too far into the wilderness of misinformation. Any errors which remain are my own responsibility entirely; I should be grateful if they were called to my attention.

This book is intended for post-graduate and honours students in biology and medicine who are learning the techniques of transmission electron microscopy, and for those workers in the biological sciences who wish to use the electron microscope as a research tool but who do not necessarily intend to become career electron microscopists. A chapter has been included on routine maintenance of the microscope which may be helpful to technicians newly in charge of microscope maintenance, or who are preparing for examinations in electron microscopy such as that given by the Institute of Science Technologists.

The scope of the present volume is largely restricted to the handling of tissue prepared for electron microscopy by embedding and sectioning. Only passing references are made to methods of studying viruses, bacteria and other particulate matter. Brief synopses of freezing techniques, high voltage electron microscopy and methods for the study of surfaces such as scanning electron microscopy are given in Chapter 10.

The references for further reading at the end of each chapter direct attention to books and papers which deal in detail with the subjects discussed in that chapter. Both early 'germinal' papers and up-to-date methods papers and review articles are cited. These are arranged, when possible, under appropriate headings.

The Appendix gives detailed instructions for some of the more useful preparative techniques, and includes a list of suppliers of the necessary chemicals and consumable goods. Attention is called to special conditions for storage of these materials where necessary.

Within the narrow confines set forth above, it is hoped that the volume will provide some help where it is needed.

1972 BRENDA S. WEAKLEY

Acknowledgements

Many people have helped me in the task of sifting fact from fantasy by reading and commenting on portions of the manuscript, and I wish gratefully to acknowledge them here: Professor R. Barer, Dr P. R. Lewis, Dr David Knight, Professor D. A. T. Dick, Dr P. J. Stoward, Dr Gladys A. Harrison, Mrs E. M. Lloyd-Davies, Mr Alan S. Pyper, Mr S. C. Turner, Mrs E. Dick and Dr J. R. Majer; also Messrs L. Slater, C. Mayo, J. Chandler and R. Partridge of Associated Electrical Industries and Mr K. O. Wood and colleagues at the Electron Probe Division of Cambridge Scientific Instruments Ltd.

Mrs E. M. Lloyd-Davies, who has assisted me for the past four years in the teaching of electron microscopy, deserves special mention for much sound advice and invaluable technical assistance.

My husband, Dr T. J. R. Weakley, undertook the heroic task of reading the entire manuscript to weed out the Americanisms and to ensure that it made sense to the non-biologist scientist. His enthusiasm for correct English usage has been of immense help.

I am very grateful to Miss Mary Benstead who prepared the line drawings with speed and expertise. Cambridge Scientific Instruments Ltd (Electron Probe Division), the Pest Infestation Laboratory, Slough, and Dr G. Dupuoy kindly supplied micrographs and gave permission for their reproduction.

Professor G. H. Bell gave helpful advice at an early stage. Dr Audrey Glauert kindly suggested readers for two of the chapters. I also thank persons too numerous to mention who have supplied me with useful references, and the staff of the Dundee University Library who helped me to locate the more difficult ones. Portions of the manuscript were typed by Mrs Margaret Dorling, Mrs A. Lyle, Mrs Margaret Duncan and Miss Veronica Borland.

Thanks are also due to the Staff of Churchill Livingstone for their co-operation and helpfulness in the production of this book.

Contents

I

Simplified Basic Theory of Electron Microscopy

RESOLVING POWER (RESOLUTION)

Imagine two points on a specimen viewed in a microscope. If these points are moved closer and closer together, a certain critical distance of separation will be reached where the two points will appear to merge and will be seen as one point only. The resolving power (resolution) of a microscope can be defined as the smallest distance by which two points on a specimen can be separated and still be distinguished as two points.

Any point on the object (specimen) plane will produce a disc surrounded by faint rings on the image plane because of diffraction effects (see the section on diffraction, p. 8). If the centres of the discs originating from two neighbouring points are separated by a certain distance the discs are said to be resolved. The maximum theoretical resolving power of a microscope is about half the wavelength of the illuminating beam.

The wavelengths of visible light fall between 0·4 μm and 0·7 μm (4000 and 7000 Å). Therefore the best resolution which can be achieved with the light microscope is approximately 0·2 μm (2000 Å).

THE WAVE NATURE OF THE ELECTRON BEAM

It was known as early as 1924 that wave properties are associated with a beam of electrons, and that the wavelength is inversely proportional to the velocity of the electrons (de Broglie, 1924). It was soon afterward ascertained that an electron beam can be focused by passing it through a magnetic field (Busch, 1926). These two characteristics of an electron beam are the basis upon which the electron microscope was developed, and a resolution of as little as 2 Å (0·0002 μm) has been achieved in favourable circumstances.

A BASIC ELECTRON MICROSCOPE

Basically, a simple electron microscope consists of an evacuated metal cylinder within which are aligned, one under the other, a tungsten filament (the cathode), a metal plate with a central aperture (the anode), a number of magnetic lenses, a fluorescent viewing screen, and a photographic plate (Fig. 1). A current passing

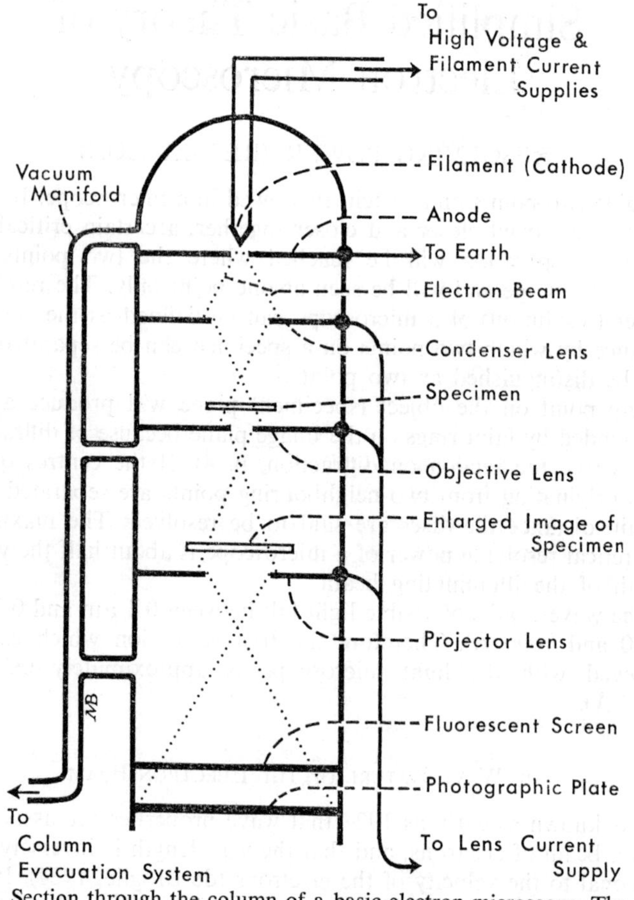

Fig. 1. Section through the column of a basic electron microscope. The path of the electron beam (dotted lines) is depicted as a classical 'ray diagram'.

though the tungsten filament causes it to heat up and emit electrons. A high negative voltage applied to the filament causes a large potential difference to occur between the filament and the anode plate, which is at earth potential. This accelerates the electrons toward the anode. Some of these pass through the central aperture of the anode and travel on down the column as an electron beam.

The electron beam is focused by the first magnetic lens (the condenser lens) so that it illuminates the specimen. Most of the electrons pass through the specimen without deviation. However, some electrons are scattered by heavy atoms in the specimen and are knocked out of the beam altogether. This forms a pattern in the emergent beam which is translated into an image when the beam is again focused. The electrons which pass through the specimen are focused by a second magnetic lens (the objective lens) which forms an enlarged image of the specimen. This image is magnified further by a third magnetic lens (the projector lens), and is viewed on the fluorescent screen. The image can then be photographed by raising the fluorescent screen so that the beam impinges upon the photographic plate. In most work micrographs are taken at magnifications of up to about 100,000 times, but these can be further enlarged photographically by a factor of up to ten times.

The voltages employed in most transmission electron microscopes to accelerate the electron beam range from 50,000 to 100,000 V (50 to 100 kV). The wavelengths associated with these voltages are 0·055 Å to 0·039 Å respectively.* However, the theoretical maximum resolution of one half this wavelength (approximately 0·02 Å) cannot in practice be attained due to a number of factors. These include difficulties in machining the parts of the microscope to the requisite perfection, instabilities in the high voltage and lens circuits, diffraction effects, and lens aberrations which make it impossible to use more than a tiny lens aperture. A resolution of 2 Å has been achieved using crystalline (periodic) specimens; 3 to 4 Å for certain non-periodic material. For most biological material, however, 10 to 20 Å is generally considered to be the approximate limit of resolution due to the techniques involved in specimen preparation and to the nature of the specimen itself. (Biological specimens are usually thin sections of plant or animal tissie supported by a metal mesh grid, or particulate matter dispersed on a thin film and supported by a grid.)

*This is calculated from the formula of de Broglie (1924):

$$\lambda = h/mv$$

where λ = wavelength
h = Planck's constant
m = mass of the electron
v = velocity of the electron

When this formula is developed mathematically and applied to an electron which has been accelerated by a potential difference, the formula becomes:

$$\lambda = \sqrt{\left(\frac{150}{V}\right)}\ \text{Å}$$

where V = the accelerating voltage.

The Electron Gun

The electron gun is composed of three elements: (1) a tungsten wire (the filament or cathode) bent in the form of a V which is surrounded by (2) a metal housing (the cathode shield) which contains an aperture located directly below the filament tip, and (3) the anode plate. The latter is a circular plate with a central aperture through which the beam of electrons can pass. The anode plate aperture is aligned so as to lie on an axis running through the filament tip and cathode shield aperture (Fig. 2).

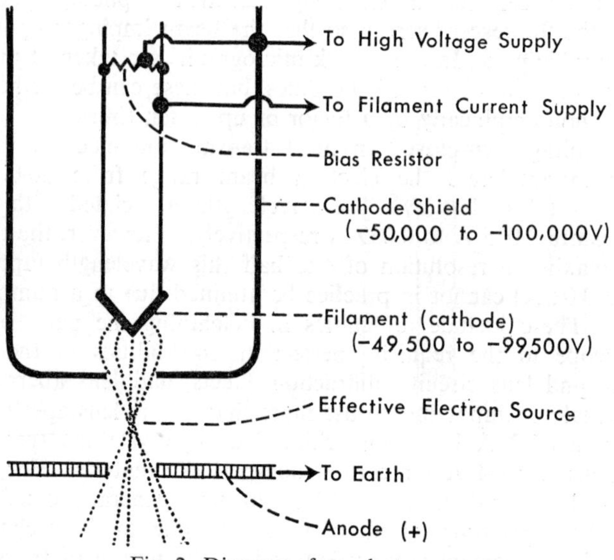

Fig. 2. Diagram of an electron gun.

The filament is energized by a heating current of a few hundred microamps. When the filament is hot enough, the metal atoms give off a cloud of electrons (thermionic emission) which tends to be more concentrated at the tip of the filament. The cathode shield is maintained several hundred volts negative with respect to the filament by inserting a resistor into the circuit between shield and filament. This negativity (known as the 'bias voltage') repels the electrons leaving the filament away from the inner surface of the cathode shield and gathers them in the vicinity of the cathode shield aperture. That is, the cathode shield itself acts as a weak lens, converging the electrons into a dense cloud approximately 40 to 50 μm in diameter. The cloud of electrons thus formed constitutes the effective electron source, and electrons from it are propelled down the column by the voltage difference between the filament and the anode plate.

MAGNETIC LENSES

A magnetic lens is basically a coil of several thousand turns of wire through which a current of one amp or less is passed. Passage of the current through the coil creates a magnetic field. Electrons are deflected by this magnetic field. To avoid the necessity for a large current passing through the coil, the magnetic field is concentrated by encasing the coil, inside and out, in a soft iron shroud in which only a small ring-shaped gap has been left facing the interior of the coil (Fig. 3). The entire field is thus concentrated within this

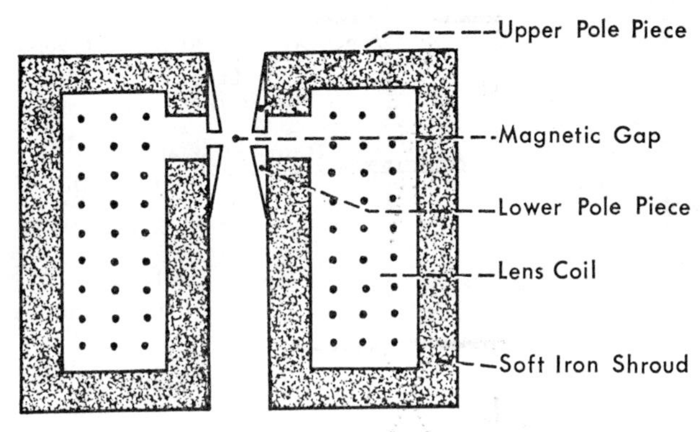

Fig. 3. Section through a magnetic lens with pole piece.

gap. To concentrate the field still further, a soft iron pole piece is inserted into the bore of the objective lens. This reduces both the bore of the lens and the width of the ring-shaped gap into which the magnetic field is now concentrated.

An electron entering the magnetic field of the lens is caused to take a helical path through the field. This is due to the summation of two forces: (1) The potential difference between cathode and anode which tends to propel the electron in a straight line down the column, and (2) The magnetic field which tends to cause the electron to assume a circular path at right angles to the electron optical axis. The net result of these two forces is a spiral, or helix.

To focus an electron beam onto a given plane the current through the coils of a magnetic lens has to be set to a very precise value. If the current is increased the beam is brought to a focus closer to in either the lens current or the electron accelerating voltage are the lens and vice versa. Fluctuations of only a few parts per million sufficient to defocus the image produced by the objective lens and limit the resolving power of the microscope.

When the focal plane of a magnetic lens in the electron microscope is changed, the visual effect seen on the viewing screen varies with the purpose of the particular lens being manipulated. The purpose of the condenser lens is to illuminate the specimen. It is a relatively weak lens with a longer focal length than that of the objective and projector lenses. The condenser lens may bring the electron beam into focus directly upon the specimen (that is, the rays of the electron beam cross over at the specimen plane, giving an intense spot of illumination of minimum diameter on a small area of the specimen (Fig. 4*a*). This is referred to as 'focusing the

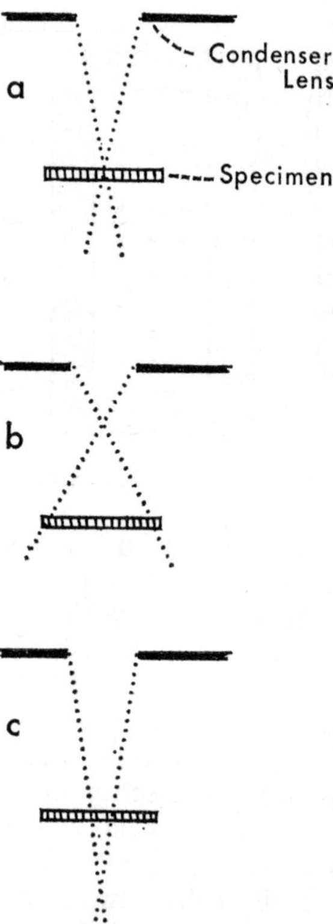

Fig. 4. Effect of 'focusing', 'overfocusing' and 'underfocusing' the condenser lens. (*a*) With electron beam focused on specimen a small intense spot of illumination is produced. (*b*) With beam focused above specimen ('overfocused') illumination is spread over specimen. (*c*) With beam focused below specimen ('underfocused') illumination is also spread over specimen.

condenser'. Alternatively, the beam may be focused either above or below the specimen plane. In either case the effect is that of spreading the beam over the specimen so that a larger area is illuminated at reduced intensity (Fig. 4*b* and *c*). Since more lens current is necessary to focus the illumination above the specimen, this is called 'overfocusing'. Since lens current is reduced to bring the illumination to focus below the image, this is called 'under-focusing'.

The objective lens is a strong lens with a highly concentrated magnetic field and a short focal length (approximately 2 to 3 mm at maximum excitation of the lens.) The objective lens causes the electron beam which has passed through the specimen to come into focus once more, at a point only a few millimetres below the specimen. A short distance below this focus point a magnified image of the specimen is produced. This image can actually be observed in those microscopes such as the Siemens which have an 'intermediate' fluorescent viewing screen inserted at the image plane. It is the quality of this image produced by the objective lens which determines the quality of the image that is finally recorded photographically. Hence it is the objective lens which determines the resolution of the microscope, and it is here that electron optical aberrations become most significant (see below).

The depth of focus of the objective lens of the electron microscope (i.e. the distance along the optical axis at which a given structure in the specimen will appear to be in true focus) is approximately 0·2 μm (2000 Å). This exceeds the depth of focus of the light microscope by a factor of two. Since the thickness of a section suitable for electron microscopy is about 600 Å, anything lying within the section will be in focus. This makes stereoscopic study of a specimen feasible, as described in the chapter on special techniques.

The size of the intermediate image (i.e. the magnification produced by the objective lens) remains constant, and is determined by the focal length of the particular objective lens. Increasing or decreasing the current in the objective lens serves only to put the image into clear visual focus. The image is then magnified by the projector lens.

The magnification produced by the projector lens is dependent upon the current passing through the coil of the lens. Increasing the current in the projector lens causes the electron beam to come to focus higher up on the optical axis. This means that the beam will be spread to a greater extent by the time it reaches the fluorescent screen. The more the beam is spread, the higher the magnification of the image at any given point along the optical axis below the lens. The projector lens has a great depth of focus (several metres) and thus the distances at which the fluorescent screen and photographic plate are placed below the lens are not critical. In other words, a screen or plate placed at any convenient place on the optical axis below the focal point of the projector lens will record an image in sharp focus, as long as the intermediate image produced by the objective lens is in sharp focus.

For simplicity the above account of microscope function has presumed a basic three-lens microscope. In practice, however, most modern transmission electron microscopes utilize two condenser lenses (referred to as C1 and C2) in order to obtain a very small area of intense illumination on the specimen for high magnification work. The function of C1 in a double condenser system is to reduce the diameter of the image of the effective electron source from 40 to 50 μm to as little as 1 μm or less ('demagnification'). The 'demagnified' image is then projected onto the specimen by C2. This makes it possible, when desired, to illuminate a very small area of the specimen, thus avoiding beam damage to other areas of the specimen. The size of the projected image of the electron source can be varied at will by increasing or decreasing the current through the C2 coil.

Two projector lenses are also commonly present (P1—sometimes called the intermediate lens—and P2) in order to attain high magnification of the final image with a reasonably short microscope column.

PHYSICAL PHENOMENA AND LENS DEFECTS WHICH LIMIT THE PERFORMANCE OF THE ELECTRON MICROSCOPE

1. *Diffraction*

Diffraction is inherent in the wave nature of both light and electron beams. It cannot be eliminated, although its effect on the image can be minimized.

Whenever waves encounter an obstacle or irregularity, new waves are generated which spread out in all directions. This process, which results in a lack of absolute sharpness in shadows, is called *diffraction.* The diffracted waves can interact with themselves and with the original wave in a complex manner giving rise to fluctuations in intensity often manifested as fringes. Image formation in both light and electron microscopes involves a double diffraction process, first at the edges or irregularities of the object and then at the edges or aperture of the objective lens. The effects of the objective lens aperture are particularly important in limiting resolution.

Because of diffraction effects, a point in the object is never imaged as a true point but as an Airy disc pattern—a central disc surrounded by a number of faint halos (diffraction fringes). The radius (d) of the central disc, in which about 80 per cent of the total energy in the pattern is concentrated, is given by

$$d = \frac{0{\cdot}61\lambda}{n \sin \alpha}$$

where λ=wavelength, n=refractive index of the medium between object and lens, and α=the semi-aperture angle of the objective lens (i.e. half the angle (θ) subtended by the edges of the lens or its limiting aperture at a point on the object, as shown in Figure 5).

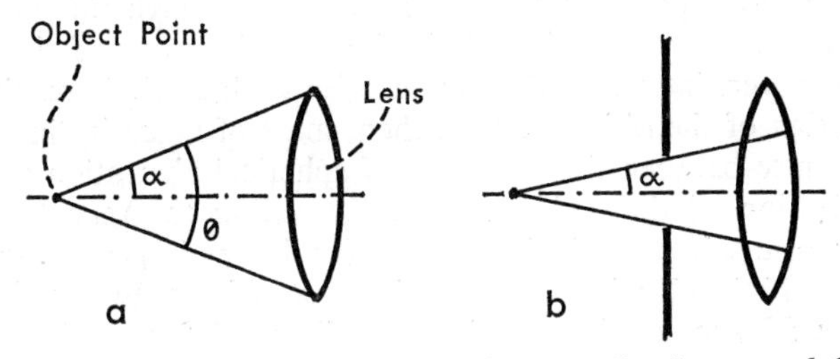

Fig. 5. The semi-aperture angle (α). (a) α limited by the diameter of the lens. (b) α limited by an aperture.

A common criterion of resolution is that if two points in the object are separated by a distance such that the centres of the Airy patterns they produce are a distance d apart they are said to be resolved. The formula for d then gives the smallest distance between two object points that is necessary for them to be distinguished as separate points.

The term 'numerical aperture' (N.A.) is used in connection with both light and electron microscopes. It is equivalent to the term $n \sin \alpha$. Since d=resolving power, the equation for resolving power is sometimes encountered in the following form:

$$\text{R.P.} = \frac{0{\cdot}61\ \lambda}{\text{N.A.}}$$

If the accelerating voltage is 60 kV the wavelength associated with the electrons will be 0·05 Å. If electron lenses were as highly developed as optical ones the radius of the central Airy disc and the minimum resolvable distance would be about half this value. In practice, various imperfections and aberrations limit the aperture angle of electron microscope lenses to about one hundredth of that obtainable in optical lenses, so that the resolution is of the order of a few ångströms (Ruska, 1966).

2. *Spherical Aberration*

This is a lens defect common to both light and electron lenses. In both cases rays traversing the peripheral part of the lens are focused onto a different plane from those near the optical axis. This leads to blurring of the image with subsequent loss in resolution. In electron microscopy this defect is particularly troublesome in the objective lens which forms and focuses the initial image of the specimen. The effects of spherical aberration can be minimized by working only with the central portion of the lens field. This is made possible by inserting a metal disc containing a small aperture into the bore of the objective lens. Only electrons which pass close enough to the optical axis of the lens to go through this aperture will contribute to formation of the image. The edges of the aperture, however, introduce diffraction effects which limit resolution. The design of the objective lens, then, must of necessity involve a compromise between the reduction of spherical aberration and the minimization of diffraction effects. According to Agar (1965) a compromise is best reached at a value for the semi-aperture angle α of about 10^{-2} rad, which is only about 36′ of arc. In practice, an aperture 50 μm in diameter is generally used for routine work; 25 μm for more critical work. The semi-aperture angle is related to aperture size as follows:

$$\alpha = \frac{\textit{Diameter of aperture}}{2f}$$

where f=the focal length of the objective lens. Example:

$$1 \times 10^{-2}\ \text{radians} = \frac{50\ \mu\text{m}}{2 \times 2500\ \mu\text{m}}$$

3. *Chromatic Aberration*

Chromatic aberration can affect resolution in electron microscopy at three levels. The first level involves the electrical stability of the microscope and is a design problem which is outside the control of the operator. The second level concerns the effect of contamination in the microscope column. The third involves the nature of the specimen being observed. The latter two levels lie to some extent within the control of the operator.

In light microscopy, chromatic aberration means that light of different colours (i.e. different wavelengths) is brought to focus on different planes, with consequent blurring of the image. In electron microscopy a similar focusing defect is encountered. This is also called 'chromatic aberration', although no colour is associated with a beam of electrons.

The wavelength of the electron beam is determined by the accelerating voltage. Therefore, any fluctuation in this voltage will of necessity cause a change in the wavelength of the electrons and thus a change in the focusing plane. To reduce chromatic effects to a level where they do not become a limiting factor in resolution, it is necessary to stabilize the accelerating voltage to the order of a few parts per million. This has been successfully accomplished in modern electron microscopes.

Chromatic aberration is also introduced because electrons do not all leave the filament at the same initial velocity. However, the chromatic aberration caused by the velocity spread of the electrons is not very great, and its effect on resolution is less than the resolution limit imposed on the microscope by the spherical aberration–diffraction compromise described above.

Chromatic aberration may also arise through the buildup of contaminating material on the inside surfaces of the column or on lens components or apertures. These contaminants may become charged sufficiently to slow down electrons in the beam and increase their wavelength. Hence cleanliness of the microscope column is essential for high resolution electron microscopy.

Perhaps the most important source of chromatic aberration as far as the operator is concerned is the specimen with which he must work. The thicker the specimen the more electrons will be scattered by it. Those electrons which are scattered sufficiently to fall outside the objective aperture are lost from the beam. This causes a pattern to be produced in the emergent beam which results in formation of an image when the beam is focused. However, some electrons which are scattered by the specimen *are not scattered strongly enough to be lost from the beam.* These electrons suffer a loss of velocity and therefore produce chromatic aberration. This results in an increase in the diameter of the disc about image points and is referred to as 'electron noise'. Electron noise is one of the factors which limits the resolution obtainable on ultrathin sections of biological material.

4. *Astigmatism*

One cause of astigmatism in electron microscopy is a non-symmetrical magnetic field in the objective lens. This results in a given specimen point being overfocused in one direction and underfocused in the other. In other words, the point cannot be brought into true focus, and the disc about the image points is enlarged.

The distribution of Fresnel (diffraction) fringes around such a point causes the point to appear elongated in one direction. Astigmatism in the objective lens will, of course, adversely affect resolution.

Asymmetry in the objective lens field is due to the following causes:

(1) Inhomogeneities in the metal used in manufacturing the objective polepiece.

(2) Impossibility of machining the polepiece so that its bore is perfectly circular.

(3) Contaminating material which becomes charged and affects the lens field. This contamination may be on the objective aperture, specimen holder, specimen, polepiece bore, or other surfaces exposed to the beam in the vicinity of the specimen.

The first two causes of astigmatism cannot be eliminated, but they can be compensated for by use of an astigmatism corrector (sometimes referred to as a 'stigmator'). An astigmatism corrector is a device which causes an equal asymmetry in the lens field in a direction perpendicular to the natural asymmetry of the lens. Thereby the effect of lens field asymmetry can be cancelled out. Detailed instruction on the correction of astigmatism of the objective lens is given in Chapter 5. If a resolution of better than 100 Å is to be attained, this correction is imperative.

The third cause of astigmatism can be eliminated by scrupulous cleanliness of the objective lens-specimen area. Astigmatism caused by charged contaminants can be compensated for by the astigmatism corrector if the astigmatism is not too severe. If the contamination is too great, however, its effects cannot be compensated for.

Astigmatism also occurs in the condenser lens, but it affects the illumination rather than the resolution. If the condenser lens is astigmatic the focused illuminating spot will appear elongate rather than round. When expanded (defocused), the beam will not illuminate the specimen evenly. A condenser astigmatism corrector is provided to compensate for this (see Chapter 5). Inherent astigmatism in the condenser lens is usually less important than the condition of the condenser aperture. If this is not circular or is contaminated, the focused beam will not be circular.

Astigmatism in the projector lenses does not materially affect image quality and need not be compensated for.

5. *Distortion*

Distortion is important mainly with respect to the projector lenses. Because electrons traversing the periphery of the lens field are affected by the field in a different manner than those passing close to the axis, the image produced by the projector lens may vary in size with distance from the axis of the lens. Particularly at low magnifications, where a large portion of the image is viewed, the periphery of the image may be magnified more than the centre, giving the effect known as 'pincushion distortion' (Fig. 6*b*). The

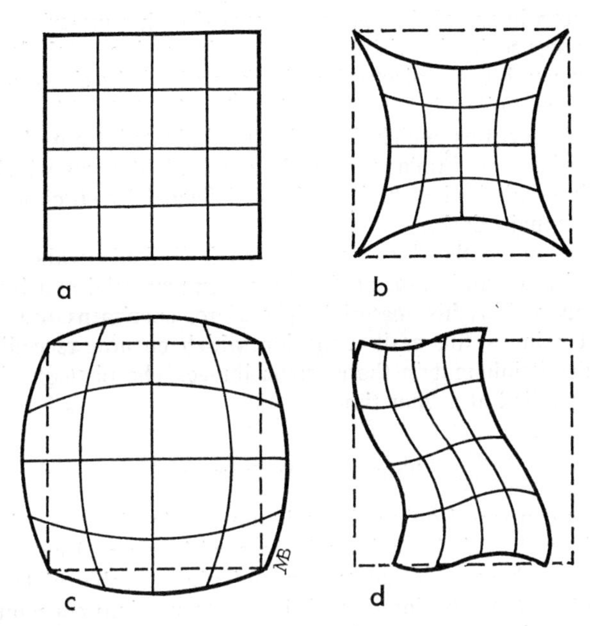

Fig. 6. Distortion. (*a*) Image free from distortion. (*b*) Pincushion distortion. (*c*) Barrel distortion. (*d*) Spiral distortion.

reverse situation also occurs: the centre of the image may be magnified more than the periphery, giving rise to 'barrel distortion' (Fig. 6*c*). These distortions can be partly compensated for by adjustment of field strength and direction in the two projector lenses.

Differences in rotation of the image in the magnetic field from optical axis to periphery can give rise to spiral distortion (Fig. 6*d*). This can be compensated for by winding the wiring of the objective and projector lenses in opposite directions so that spiral distortion produced by one lens can be cancelled out by that produced by the other.

Because of distortion it is seldom possible to use magnifications below 1000 for serious study of a specimen, although lower magnifications are useful for locating an area for study.

The Fluorescent Screen

Fluorescence is the property of emitting electromagnetic radiation under the influence of electromagnetic or electron beam bombardment. Certain materials emit visible light under these conditions.

This phenomenon is useful in electron microscopy for making the electron image visible. A screen coated with a material capable of fluorescence in the visible range, such as zinc sulphide, is installed beneath the projector lens in the path of the electron beam. The screen emits visible light when bombarded with electrons. Deflected electrons which have been eliminated from the beam cannot, of course, impinge upon the screen. The areas which they would have hit if they had not been deflected by specimen components remain dark.

Resolution on the fluorescent screen is limited to about 70 to 100 μm by the grain size of the fluorescent material and by light-scattering within this material. Therefore, specimens are photographed using a fine-grain emulsion which ensures recording all of the detail inherent in the electron image. The photographs can then be studied at a later time.

The Vacuum System

Electrons can travel only a few micrometres in air before they are stopped or slowed down by collisions with gas molecules. Since the distance between electron gun and photographic plate is approximately 1 m, the interior of the electron microscope must be evacuated. The usual operating vacuum is 10^{-4} mmHg,* in which an electron can travel, on the average, $2\frac{1}{2}$ m before encountering a gas molecule.

There are also other reasons for evacuating the column. If gas molecules are present in the space between filament and anode they are liable to be converted into positively charged ions by bombardment from the electron beam. These ions can increase the conductivity of the filament–anode gap sufficiently to allow discontinuous electric discharges to occur between filament and anode instead of a steady beam of electrons.

Filament life is decreased by operation of the microscope in a

* 1 mmHg = 1 millimetre of mercury = the pressure exerted by a column of mercury 1 mm high at 0°C. This unit is also called 1 Torr (in honour of Torricelli). The equivalent to 1 mmHg in SI units (see p. 16) is 1·33 millibars (mbar). Pressures higher than 10^{-3} Torr are considered to be in the low-vacuum range; pressures lower than this are considered to be in the high-vacuum range.

poor vacuum. Bombardment by positive ions and oxidation of the tungsten both weaken the filament. Oxidation also reduces the efficiency with which the filament can emit electrons.

In order to maintain the requisite vacuum, two types of vacuum pump are necessary. A rotary mechanical fore-pump (the 'roughing pump' or 'backing pump') first 'rough pumps' the column. This pump is located external to the microscope to avoid vibration but is connected with it by a vacuum duct. This pump can only maintain a vacuum in the order of 10^{-2} mmHg. When this vacuum has been reached, a second type of pump, the diffusion pump, takes over and increases the vacuum to 10^{-4} mmHg. The diffusion pump can work efficiently only when the initial gas pressure has been reduced by the fore-pump. The diffusion pump is incorporated into the microscope itself.

A diffusion pump may be operated with either oil or mercury, but oil is more commonly used. Oil vapour is produced by a heater within the diffusion pump. Air molecules are trapped within a stream of this vapour which is conducted away from the heater and condensed in a cooler region of the pump. The air molecules released at condensation are removed by the forepump (which, under these circumstances, is commonly referred to as the 'backing pump', since it backs up the action of the diffusion pump). The condensed oil returns to the heaters to repeat the cycle.

The vacuum attainable by an oil diffusion pump is determined by the vapour pressure exerted by the oil. Use of a high-vapour-pressure oil would result in contamination of the column with oil vapour molecules. It is thus necessary to use a special high-vacuum oil of low vapour pressure.

So that the entire column need not be returned to atmospheric pressure each time the specimen or photographic plates are removed or inserted, special air locks are built into the column at appropriate places. These can be closed off from the rest of the column when required, and permit the air admitted to be confined to the specimen chamber or to the camera chamber as the case may be. These small chambers can then be returned rapidly to a vacuum of 10^{-2} mmHg by opening the valves which connect them directly to the forepump. The air locks can then be opened to the column with little loss of vacuum to the latter.

The Cooling System

The diffusion pump and the magnetic lenses are kept cool during operation of the microscope by water circulating through special cooling jackets. If unfiltered tap water were used for this purpose, matter in suspension would soon block the cooling system. As an alternative to frequent changing and cleaning of filters, many microscope owners install a recirculation system employing deionized or distilled water.

UNITS OF LENGTH AND PRESSURE IN ELECTRON MICROSCOPY

The Système Internationale d'Unités (SI) for units of measurement was advocated in 1960. In SI, the units of the basic quantities length, mass and time are respectively the metre (m), kilogram (kg) and second (s), and the units for derived quantities are stated in terms of these base units. For instance, pressure has the dimensions of force per unit area, i.e. of mass/(length × time^2) and the SI unit of pressure is 1 kg m^{-1} s^{-2}. The tables below give the non-SI units of length and of pressure (the quantities most frequently encountered by electron microscopists), and their SI equivalents.

At the time of writing, certain journals specializing in electron microscopical studies such as the *Journal of Ultrastructure Research* (U.S.), *Journal of Cell Biology* (U.S.), *Journal de Microscopie* (France) and *Zeitschrift für Zellforschung und Mikroskopische Anatomie* (Germany) have not changed to the SI system. The reader is therefore advised to become thoroughly familiar with both systems.*

Table 1. *Units of length*

Old system	SI equivalent
1 millimetre (mm) = 10^{-3} metre	1 millimetre (mm)
1 micron (u) = 10^{-6} metre	1 micrometer (μm)
1 millimicron (mμ) = 10^{-9} metre	1 nanometre (nm)
1 ångström unit (Å) = 10^{-10} metre	0·1 nanometre

Table 2. *Units of pressure*

Old system	SI equivalent
1 atmosphere = 760 mm mercury (mmHg) = 760 Torr = 1·013 bar	1·013 × 10^5 kg m^{-1} s^{-2} = 1·013 bar
1 bar = 0·987 atmospheres	1 bar = 1 × 10^5 kg m^{-1} s^{-2}
1 mmHg = 1 Torr = 1·33 millibars (mbar)	133 kg m^{-1} s^{-2} = 1·33 mbar

*It appears that ångström units will be used indefinitely in electron microscopy.

Further Reading

The foregoing presentation of the basic theory of the electron microscope is given in its simplest terms for the benefit of those readers with a minimal knowledge of physics and mathematics. For further reading at a more advanced level, the references at the end of this chapter, especially Sjöstrand (1967), and Meek (1970) are recommended. Excellent histories of the development of the electron microscope are given by Bradbury (1967) and Marton (1968).

REFERENCES

Agar, A. W. (1965). In *Techniques for Electron Microscopy*, Chap. 1 (Edited by D. H. Kay), Oxford: Blackwell Scientific Publications.

Bradbury, S. (1967). *The Evolution of the Microscope*, Chap. 8, Oxford and New York: Pergamon Press.

Broglie, L. de (1924). A tentative theory of light quanta. *Philosophical Magazine* **47**, 446–458.

Busch, H. (1926). Calculation of trajectory of cathode rays in axially symmetric electromagnetic fields. *Annalen der Physik* **81** (Ser. 4), 974–993.

Cosslett, V. E. (1951). *Practical Electron Microscopy*, London: Butterworth.

Grivet, P. (1965). *Electron Optics*, Oxford: Pergamon Press.

Haine, M. E. (1954). In *Advances in Electronics and Electron Physics*, Vol. 6, New York: Academic Press.

Haine, M. E. & Cosslett, V. E. (1961). *The Electron Microscope: The Present State of the Art*, London: E. & F. N. Spon.

Hall, C. E. (1966). *Introduction to Electron Microscopy*, 2nd edn, New York: McGraw-Hill.

Marton, L. (1968). *Early History of the Electron Microscope*, San Francisco: San Francisco Press.

Meek, G. A. (1970). Practical Electron Microscopy for Biologists, Chap. 1, 3 and 5, London and New York: Wiley-Interscience.

Ruska, E. (1966). In *Advances in Optical and Electron Microscopy*, London and New York: Academic Press.

Siegel, B. M. (1964). *Modern Developments in Electron Microscopy*, Chap. 1, London: Academic Press.

Sjöstrand, F. S. (1967). *Electron Microscopy of Cells and Tissues*, Vol. 1, Chap. 2, 3 and 4, New York and London: Academic Press.

Zworykin, V. K., Morton, G. A., Ramberg, E. G., Hillier, J. & Vance, A. (1945). *Electron Optics and the Electron Microscope*, New York: John Wiley.

2

The Processing of Tissues

CAUTIONARY NOTE

Several of the substances used in the preparation of tissues for electron microscopy are toxic and must be handled with care:

Osmium tetroxide. Osmium tetroxide should be used only in a fume cupboard. It is poisonous and highly volatile. The vapour irritates the entire respiratory tract and may severely burn the eyes. The solution is corrosive and causes burns. If splashed on the skin, wash at once with running water.

Glutaraldehyde. Glutaraldehyde is not as volatile nor as poisonous as osmium tetroxide, but breathing its vapours is not recommended. The solution may cause dermatitis if allowed to wet the skin.

Epoxypropane (*propylene oxide*). Epoxypropane is also poisonous, highly volatile, and should be used only in a fume cupboard. *It is also highly inflammable and may explode if used in the same room with an open flame.* It should be stored in a refrigerator.

Acetone and xylene. Acetone and xylene are highly inflammable and should not be used near an open flame. If inhaled over a period of time they may cause liver damage.

Epoxy resin embedding media (*Araldite, Epon and their catalysts and hardeners*). These materials are all toxic, and may cause dermatitis in susceptible individuals. Epoxy resins have proved carcinogenic in laboratory animals. The use of gloves while handling embedding materials is recommended.

The methacrylates. The vapours may be harmful; it is advisable to use these materials in a fume cupboard. The non-water-soluble methacrylates are inflammable, and care should be taken during pre-polymerization procedures.

Benzoyl peroxide (*dibenzoyl peroxide* (*'Lucidol'*)). This is used as a catalyst with the methacrylates. It decomposes violently at 106°C but is relatively safe at normal temperatures. It frequently comes from the manufacturer as a wet paste. It is advisable to store it in a refrigerator.

Luperco C.D.B. This is sometimes used as a catalyst with glycol methacrylate. It is 50 per cent 2, 4–dichlorobenzoyl peroxide in dibutyl phthalate. It is more unstable and hence more dangerous to use than benzoyl peroxide. It apparently has no particular advantage over benzoyl peroxide and therefore its use is not recommended.

Amyl acetate. This is used as a solvent for collodion in preparing coated specimen-support grids. It is an anaesthetic and should be used only in a well-ventilated room. Inhalation of its fumes over a period of time may result in liver damage.

Lead. Lead from lead-containing 'electron stains' can be absorbed through the skin. If stain is spilled and allowed to dry, dust containing lead may be inhaled. Always sponge off the bench before and after staining, and wash hands. Cumulative lead poisoning can lead to sterility or death.

Photographic developer. Photographic developer may cause dermatitis in susceptible individuals.

CHOICE OF ANIMAL AND REMOVAL OF TISSUES

Other factors being equal, it is an advantage in electron microscopy to use a small animal. A hamster ovary measuring a few millimetres across will yield an adequate sample of sections far more quickly than the ovary of a cow, which is the size of a small fist.

The chance of viewing completely normal tissue ultrastructure in any animal is, of course, a matter of statistical probability. The probability is increased if the animal is young and apparently healthy.

In choosing the method of killing the animal, it should be remembered that *post mortem* changes commence at death. The rate of disorganization varies from one tissue to another. In practice it has been found to be very difficult to sort out *post mortem* changes which occur before the tissue has been stabilized by the fixing fluid from changes which occur during the subsequent tissue processing. Trump and Ericsson (1965) have pointed out that actual *post mortem* (autolytic) breakdown of cell organelles is probably preceded by a state of increased fragility which would predispose them to excessive processing damage. More investigations need to be made in this field, and workers who have published on the subject disagree about the importance of speed in tissue removal. While the controversy remains unresolved, however, it seems reasonable to get the tissue out of the animal and into the fixative before the animal has died. The animal may be injected with an anaesthetic dose of Nembutal, and the tissue exposed as soon as the animal is insensate. If feasible, fixation is started before removing the tissue, by dropping fixative onto the organ *in situ*. The tissue is then removed as quickly as possible to a drop of fresh fixative, on a firm surface. Then it is diced into cubes, not exceeding one half to one millimetre, and placed in a small bottle (carefully labelled!) containing a few millilitres of the fixative. Transfer of the tissue can be easily made with a wide-mouthed Pasteur pipette. It is essential that the blocks of tissue do not exceed one cubic millimetre in size, otherwise the interior will be inadequately fixed or unfixed.

Once the tissue has been removed, the animal is killed by a further injection of Nembutal.

Although this method permits the removal of tissues while the animal is still alive, it has not been determined to what extent the nembutal anaesthesia may alter the ultrastructure of the various tissues. Since under the existing Cruelty to Animals Act tissues cannot be removed from a living unanaesthetized animal, a comparison of such tissues with those removed from anaesthetized or dead animals cannot be legally made in the United Kingdom. Even if such a comparison were made, the stress to a living unanaesthetized animal might be such as to alter the tissue ultrastructure in unpredictable ways. Comparison can be made, then, only between tissues removed from living animals while under a variety of anaesthetics (such as ether, chloroform, or Nembutal) and tissues removed from animals which have been killed by one of these anaesthetics, or by such physical means as a blow on the head. If such a study has been done at the ultrastructural level it has not come to my attention. At the light microscopical level, however, Al-Azzawi and Stoward (1970) have shown a difference in both quantity and quality of reactions designed to show the presence and distribution of lysosomes in muscle tissue, depending upon the method of killing used.

FIXATION

The purposes of fixation are (1) to halt *post mortem* changes, and (2) to preserve the tissues in a condition which resembles the living state as closely as possible.

For the beginner, one of the most common reasons for failure to attain proper fixation is the size of the block of tissue that he attempts to fix. 'One cubic millimetre' and 'one half cubic millimetre' have little meaning unless cubes of that size have been actually observed. We keep two small glass bottles containing coloured plastic cubes of these dimensions on view in the laboratory for use as a standard.

Dicing the tissue into cubes, even with a very sharp razor blade, will cause damage to the outer layers of cells; time taken for fixative penetration may mean less than adequate preservation of the block interior. With osmium fixatives a narrow zone of optimum fixation is usually found just below the surface of the blocks (Palade, 1952).

It was originally believed that the aldehydes, being light molecules, penetrated tissues more rapidly than osmium tetroxide and that therefore larger blocks of tissues could be successfully fixed. However, experiments by Ericsson and Biberfeld (1967) indicate that the penetration rate of glutaraldehyde into tissues is actually *slower* than that of osmium tetroxide. They conclude that optimum preservation throughout the block is not attained with the fixative if the size of the block exceeds one half cubic millimetre.

There is disagreement about the optimum temperature for fixation. It has long been common practice to fix at O°C in order to reduce *post mortem* changes and extraction of tissue components to a minimum. However, reduction in temperature also slows down the rate of fixative penetration. How nearly these factors cancel each other out in practice has not been accurately determined, and varies with the fixative. For further discussion see Hayat (1970).

Time is another variable. Two to 4 hours in aldehyde followed by 1 hour in osmium tetroxide is common practice.

The 1 hour fixation time in osmium tetroxide for blocks one cubic millimetre or less in size was determined empirically. It is essentially a compromise between the need for complete penetration of the fixative into the block and the fact that protein and other substances are progressively leached out of the tissues when they are immersed in osmium fixative (Bahr, 1955). Burkl and Schiechl (1968) have reported a penetration rate for osmium tetroxide solutions into tissues of 0·75 mm in 1 hour. This agrees with the rate reported by Hagström and Bahr (1960) for 1 per cent OsO_4 in distilled water. Hagström and Bahr found that addition of buffer and sucrose to the fixative retards penetration, however.

One hour to several days fixation in glutaraldehyde was used by Sabatini *et al.* (1963), who introduced this fixative. The optimum time cited by them was one half to 2 hours. However, experiments by Ericsson and Biberfeld (1967) and Hopwood (1967) suggest that 4 hours for a block measuring 1 mm^3 is a more reasonable time. Timings for aldehyde fixation as short as 15 min have been used for special cytological techniques, but ultrastructural preservation is poor. Once in contact with tissue elements, glutaraldehyde fixes rapidly. Formaldehyde, on the other hand, penetrates rapidly, but the reactions which occur during fixation take a number of hours to complete.

The optimum time for the fixation of any tissue varies with tissue type, age, block size, fixative concentration, and buffer used. If the bottles containing the tissue are placed on a turntable and revolved slowly during fixation and dehydration, penetration of fixatives and dehydrants is facilitated. Presumably the speed of extraction processes may be enhanced as well, however.

The pH is usually held at physiological levels during fixation (pH 7·3 to 7·4). Schultz and Karlson (1965) report that there is little difference in the appearance of tissues when fixed between pH 6 and pH 8. Others (e.g. Hayat, 1970) stress the importance of maintaining a physiological pH, particularly in high resolution work.

As mentioned earlier, fixation may be commenced before the tissue is removed from the animal. If the tissue to be fixed has a capsule, this may have to be slashed to allow penetration of the fixative, and any tissue fluid or blood flushed away with excess fixative.

Perfusion of the fixative into the animal through the circulatory system has been recommended as the most efficient way of getting the fixative into the interior of the tissue. Usually a cannula is inserted into the main blood vessel leading to the tissue, and secured. Then the main vein leading from the tissue is severed. The blood is flushed out of the organ with excess fixative; then the exit vein is clamped off and a slight pressure of fixative allowed to build up to dilate the vessels and allow fixative to pass through them into the tissues. The pressure must not, of course, be excessive or the tissue will be disrupted. After 5 min or so the tissue may be removed, cut into 1 mm cubes, and further fixed by immersion in fresh fixative.

Osmium tetroxide is not recommended for perfusion since it acts as a vasoconstrictor and little if any of the fixative will manage to leave the vessels and enter the tissues. The aldehydes are much more successful. Perfusion has proved of particular value in such tissues as brain and kidney which are notably difficult to fix properly. However, perfusion is generally thought to be too time consuming to be used for routine purposes on less intractable tissues. For details of this method see Pease (1964), Karlsson and Schultz (1965), Maunsbach (1964 and 1966*a* and *b*) and Schultz and Case (1970).

MECHANISMS OF FIXATION

The fixatives in common use for electron microscopy (osmium tetroxide, the aldehydes, and potassium permanganate) are generally used in aqueous solution. This solution is held at a physiological pH by the use of a buffer, and at a physiological tonicity by the use of some osmotically active substance. In considering what happens to the tissue during the process of fixation, the action of all of these substances within the fixative solution must be considered.

Although a few reactions that may occur between the fixatives and the cell constituents have been determined, many more are still obscure. Moreover, the manner in which the addition of various buffers and osmotic agents to the fixative may alter these mechanisms is not clearly understood. The picture is still fragmentary, and much research yet remains to be done. The following is a brief account of some of the more generally accepted mechanisms of fixation.

Both osmium tetroxide and the aldehydes act as cross-linking agents. That is, by the formation of chemical bonds between molecules of cell substance they produce a stable network throughout the cell. This prevents the loss of the material involved in the network itself. Secondarily, this may prevent or impede the loss of material caught within the network.

Other types of reactions can and do occur, and the reactive groups differ with the fixative used. Bonds can be formed either between the fixative molecules and tissue components, or between reactive groups in the tissue which have been unmasked by the fixative (Millonig and Marinozzi, 1968; Pearse, 1968).

Osmium Tetroxide

Osmium tetroxide is known to react with double bonds in lipid (Bahr, 1954). It is thought to cross-link neighbouring molecules in the following manner (Stoeckenius and Mahr, 1965; Korn, 1967):

```
 |                         |           |              |
 CH    O     O            CH         H-C-O         O-C-H
 ||     \\  //            ||           |    \      /
 ||  +    Os       +      ||   --->    |      Os            +H2O
 ||     //  \\            ||           |    /  ||  \
 CH    O     O            CH         H-C-O     ||   O-C-H
 |                         |           |       O      |
```

Osmium tetroxide is also thought to react with the polar groups of lipids after its initial reaction with the double bonds (Riemersma, 1968).

The reactions of osmium tetroxide with protein are still in doubt. It appears to cross-link protein (Hake, 1965). SH and SS groups are thought to react (Bahr, 1954, 1955), as are the side-chain amino groups (Sabatini *et al.*, 1963), but the validity of these assumptions has been questioned (for discussions see Pearse, 1968; Hayat, 1970).

Nucleic acids and carbohydrates do not appear to react with osmium tetroxide. If nucleic acid is bound to protein, however, as in ribosomes and chromosomes, it is retained by fixation of the protein portion. Glycogen aggregates, being relatively insoluble, are usually held within the tissue by the network of fixed protein. Small soluble molecules are usually lost.

The Aldehyes

Neither unsaturated lipids nor phospholipids are stabilized by aldehyde fixation. Unless further 'post fixation' in osmium tetroxide is carried out, these materials will be dissolved in the dehydration liquids and lost. Aldehydes are however more efficient cross linking agents for protein than osmium tetroxide.

As pointed out by Pearse (1968) the reactions of formaldehyde and protein are numerous and complex. However the following two-step cross-linking reaction between formaldehyde and side-chain amino groups is of particular interest to electron microscopists:

Step 1

$$\Big| -NH_2 + HCHO \longrightarrow \Big| -\overset{\displaystyle H}{\overset{|}{N}}-CH_2(OH)$$

This product can now react with a second protein molecule by the formation of a 'methylene bridge':

Step 2

$$\Big| -\overset{\displaystyle H}{\overset{|}{N}}-CH_2(OH) + H-\overset{\displaystyle H}{\overset{|}{N}}- \Big| \qquad \Big| -\overset{\displaystyle H}{\overset{|}{N}}-CH_2-\overset{\displaystyle H}{\overset{|}{N}}- \Big| + H_2O$$

The reactions of dialdehydes (e.g. glutaraldehyde) with protein molecules are also numerous and complex. A simple direct linkage of a monomeric glutaraldehyde molecule between two protein chains is no longer thought to be likely. Richards and Knowles

(1968) have found that glutaraldehyde in aqueous solution consists of monomers, dimers, trimers and even higher polymeric forms. It also contains significant amounts of unsaturated aldehydes resulting from aldol condensations within the solution. Any of these molecular species could contribute to the fixation process and this may be the reason for the remarkable cross-linking power of this fixative. Richards and Knowles (1968) suggest that the stable and irreversible cross linkages which occur are most likely due to reactions with the unsaturated aldehydes.

The cytoplasm is far better preserved by glutaraldehyde than by osmium tetroxide alone. Such organelles as microtubules, mitotic spindle fibres and cytoplasmic lamellae in certain rodent eggs have been routinely seen only after the introduction of this fixative. Glutaraldehyde is superior to the other aldehydes for the preservation of fine structure and is therefore the aldehyde of choice for most work.

Formaldehyde is a less efficient protein fixative than glutaraldehyde but is often used when preservation of enzyme activity in the tissue is of primary importance. Karnovsky (1965) and Friend and Farquhar (1967) recommend a mixture of glutaraldehyde and formaldehyde, reasoning that the rapidly penetrating formaldehyde will halt *post mortem* changes, and further cross-linking will be accomplished when the glutaraldehyde arrives. Schultz and Case (1970) question the advisability of using formaldehyde at all, even when combined with glutaraldehyde, if optimum morphological preservation is to be attained.

The use of formaldehyde and other aldehydes in enzyme histochemistry will be discussed in Chapter 8.

The purity of the aldehydes is important if the quality of fixation is not to suffer. Formaldehyde fixatives should be made fresh from paraformaldehyde powder; glutaraldehyde fixatives from specially stabilized 'E.M. grade' solutions. For details, see Appendix, pp. 197–199.

Double Fixation

By far the most popular method of tissue preservation at the present time is 'double fixation'—i.e. primary fixation in glutaraldehyde followed by 'post fixation' in osmium tetroxide. This method combines the advantages of both fixatives. More protein is retained than with osmium tetroxide alone, and some neutral and most phospholipids are stabilized by the osmium tetroxide.

Permanganate

Permanganate fixatives, usually in the form of potassium permanganate, have been widely used for the fixation of plant tissues, and to a lesser extent for animal tissues. These fixatives have now been largely superseded by glutaraldehyde.

The chemistry of fixation with permanganate is not well understood. It is an oxidizing agent, as is osmium tetroxide, but its effects on tissues are considerably different than those of osmium tetroxide. Bradbury and Meek (1960) found that RNA and histones are removed during permanganate fixation. They suggested that potassium permanganate does not act as a true fixative at all, and that what fixation is accomplished is done by ethanol during subsequent dehydration. Experiments *in vitro* by Hake (1965) indicate that whereas osmium tetroxide cross-links proteins and stabilizes them, potassium permanganate solubilizes certain proteins and leaves the rest drastically altered. During processing of the tissues a large part of the cellular protein and lipid are lost, and mitochondria appear swollen. Some DNA is retained, but clumped chromatin is not visible in the nucleus. Ribosomes are not seen. However, phospholipid membranes are very well preserved, and studies with permanganate-fixed tissues have made a considerable contribution to the knowledge of membrane structure (e.g. Sjöstrand, 1962).

For reviews of the literature on permanganate and other metal-containing fixatives see Trump and Ericsson (1965), Hopwood (1969), and Hayat (1970).

Plate 1 illustrates the typical appearance of tissues fixed in three different fixatives.

Recipes for commonly used fixatives will be found in the Appendix.

Buffers Commonly Used With Fixatives

Veronal Acetate Buffer

This buffer achieved wide popularity since it was the one recommended by Palade (1952) in his original osmium tetroxide fixative. It cannot, however, be used with aldehydes, since a reaction occurs between the fixative and the buffer which destroys its buffering capacity.

Phosphate Buffers

These have become increasingly popular for both osmium and aldehyde fixatives. They are so-called 'physiological buffers' since they are present in living systems, they are non-toxic, and their buffering capacity at pH 7·4 is strong. Millonig's phosphate buffer is particularly useful with osmium fixatives as it helps prevent leaching of protein and other material during fixation (Millonig, 1962).

Cacodylate Buffer
Sodium cacodylate buffer was used by Sabatini *et al.* (1963) with glutaraldehyde and is still in common use. It is effective with osmium tetroxide as well as with the aldehydes.

s-Collidine
Bennett and Luft (1959) introduced *s*-collidine as a buffer for fixatives. They found that use of this buffer increases the ease of sectioning the tissue. It also has the advantage of remaining stable and contamination-free indefinitely. It was slow in coming into common use because the necessary degree of purity was difficult to obtain commercially and procedures for purifying it in the laboratory are tedious. However, it is now available in a special 'E.M. grade' from most suppliers of materials for electron microscopy.

Recipes for several buffers are given in the Appendix.

LOSS OF TISSUE COMPONENTS DURING PROCESSING

A number of attempts have been made to determine what materials are lost or retained during fixation and dehydration procedures. Several methods have been used on a variety of tissues, and a wide range of values for loss of various materials has been reported. Extent of loss depends upon a number of factors, including tissue type, fixative, fixation time, and buffers used during fixation and rinsing. For reviews of the literature see Hopwood (1969) and Williams (1969).

Unfortunately, it may well turn out to be the most important components of a cell from the physiological point of view which are washed away. Many of these are soluble. Diffusible ions such as Na^+ may be almost entirely lost. The effect of fixatives on electrostatically bound ions is almost completely unknown. The static image of the cell which we observe in the final micrograph is at best only a poor reflection of cell dynamics.

Material may also be lost during infiltration and embedding, although as yet few papers have probed this aspect.

Table 3 attempts to summarize, in very general terms, some of the information available.

Table 3. Material lost from tissues during fixation and dehydration procedures

	OsO_4	Glutaraldehyde	Formaldehyde	$KMnO_4$
Protein	Loss varies with buffer used; may be as high as 50%	Forms more cross links than OsO_4 or formaldehyde, buffer less important. Nevertheless some loss occurs	Loss varies with tissue, 4 to 12% reported in one paper	Mostly lost from tissue
Unsaturated lipids	Partial loss	Some retained by fixative but lost during dehydration	Some retained by fixative but then lost during dehydration	Mostly lost during dehydration
Saturated lipids	Mostly lost	Mostly lost	Mostly lost	Mostly lost
Phospholipids	10 to 30% loss has been reported	Retained during fixation but lost during dehydration	Some retained by fixative but lost during dehydration	Well fixed; phospho lipid–protein complexes are unmasked. About 25% lost during dehydration
Glycogen	Preserved with all buffers, best with Millonig's buffer	Approximately 75% retained in liver	Approximately 75% retained in liver	Preserved in 100 to 150 Å particles
Mucopoly saccharides	Combined glutaraldehyde–OSO_4 fixation retains hyaluronic acid; no other information available		Mostly lost	Lost
DNA	Mostly retained if in combination with protein	Mostly retained if in combination with protein		Some loss; authors disagree
RNA	Probably also retained by virtue of combination with protein; figures not available			Probably lost; ribosomes not visible

OSMOLARITY

Since osmolarity is a term which keeps appearing in the literature of electron microscopy, a brief discussion of this term may be helpful. A few pertinent definitions follow:

One mole—The number of grams of a substance which is numerically equivalent to its molecular weight. For example sodium hydroxide (NaOH) has a molecular weight of 40 atomic mass units, therefore 1 mole equals 40 g.

A molar solution—A solution containing 1 mole of solute per litre. For example 40 grams of NaOH dissolved in enough water to make up 1 litre.

A molal solution—One mole of solute added to 1 kg of solvent. When the solution is very dilute the volume of the solute is negligible. Therefore, since a kilogram of water occupies 1 litre, *molal* concentration is considered to equal *molar* concentration for dilute aqueous solutions.

The molarity of a substance in solution=the molar concentration (M)=grams per litre divided by molecular weight.

The total molarity of a solution=the sum of the individual molar concentrations of its solutes.

If two solutions of different molarities are separated by a semi-permeable membrane which permits the passage of water but not the passage of solute particles, water will pass across the membrane into the solution containing the highest concentration of solute until equilibrium is reached, i.e. until the molarities on each side of the membrane are the same. This is known as *osmosis*. The pressure which must be applied to the solution of higher concention to keep water from passing into it equals the difference in osmotic pressure between the two solutions. In biological terminology, the unit of osmotic pressure which is commonly used is the *osmole* or the *milliosmole*.

If the initial molarities of two solutions separated by a semi-permeable membrane are the same, no net transfer of water will occur between them, and the solutions are in equilibrium. In biological parlance they are frequently referred to as 'isosmotic'.

This relatively simple picture becomes complicated if the molecules of the substance in question dissociate to give charged particles when they go into solution, or if the substance is in any case ionic (salts). Strong acids and bases and all water-soluble salts dissociate completely, thus multiplying the number of particles present in solution by a number dependent on the chemical equation for the process. Thus 1 mole of sodium chloride (NaCl) in solution affords 1 mole of Na^+ ions and one mole of Cl^- ions, yielding a total of 2 moles of particles. However, due to electrical interactions between ions and other effects, the osmotic pressure developed by an electrolyte solution is in general lower than what would be expected from the particle concentration. For example the osomotic pressure of 0·1 M sodium chloride is 7 per cent lower than expected.

The concentration of particles=the molar concentration times the number of particles per molecule yielded by dissociation. Some authors have referred to the total concentration of particles in solution as the 'osmolarity' or the 'osmolar concentration'. However, they are making an assumption that concentration is proportional to osmotic pressure. As pointed out by Dick (1966), this assumption is erroneous owing to the interaction of molecules and ions with one another in solutions, as mentioned above.

However, the amount by which a solute will depress the freezing point of water is directly related to its osmotic pressure. One mole of non-electrolyte in 1 kg of water will depress the freezing point by 1·86°C. The osmotic pressure (osmolarity) of any aqueous solution can be determined quite accurately from its freezing point by simple proportionality. For example the osmotic pressure of a solution for which the freezing point depression is 0·279° C is (0·279/1·86)=0·15 osmol. For further discussion see Dick (1966).

Even if the osmolarity of a solution is accurately determined by the freezing point depression method, however, the osmotic effect of the solution on cells in a given tissue is not entirely predictable. Biological membranes are not true semipermeable membranes. The plasma membrane around the cell is not passive: it can to some extent pick and choose what ions and therefore how much water it takes in or expels. Once taken into the cell, material may be cut off from other cell components by enclosure within intracellular membranes, the composition of which varies from one organelle to another.

The fluids of the cytoplasm and those within the various intracellular compartments or organelles contain not only salts but proteins. The mixture of salts and proteins causes osmotic behaviour which is very different from that which occurs when electrolytes alone are in solution. Also, fluid composition may vary greatly from one cell type to another, and may fluctuate in any one cell with functional state.

In short, the biologist is faced by an extremely complex situation. In practice, he deals with it in empirical terms. He uses words like isotonic, hypertonic and hypotonic to describe the observed effects that solutions have on cells and tissues.

An isotonic solution—A solution in which a cell retains the volume it had in the living tissue.

A hypertonic solution—A solution in which a cell shrinks.

A hypotonic solution—A solution in which a cell swells.

The osmolarity of blood plasma has been determined to be approximately 0·3 osmoles, and this is isotonic for most mammalian cells. Sodium chloride 0·16 M (i.e. physiological saline, 0·9 per cent solution) has an equivalent effect. Therefore many solutions which are used to bathe living tissue have been developed, on the basis of freezing point determinations, to closely approximate the osmotic effect of blood plasma. Others, however, have simply been devised so that their total particle concentration equals 0·3 M.

Experiments designed to measure the osmolarity within cells (reviewed by Dick, 1966) suggest that the tonicity of blood plasma is a reliable standard for intracellular tonicity.

The situation is much more complicated, however, when a solution is used for the purpose of fixing a tissue. The ideal fixative should allow the cell to neither swell nor shrink during fixation. Opinion in the literature is divided as to whether it is the total concentration of the fixing solution or the concentration of the medium in which the fixative is dissolved which is important (cf. Maunsbach, 1966*b*; and Weibel, 1969). All too often the most important point is overlooked: that is, that very little is known about what happens to a cell membrane when fixative molecules interact with it. In its normal state the cell membrane is selectively permeable to particular ions. Exactly how long it maintains this selectivity after contact with the various fixatives and their media is not known. Experiments by Elbers (1966) indicate that some osmium fixatives cause the cell membranes of certain eggs to become completely permeable to ions within a few seconds. Permeability changes are slower with gluteraldehyde.

Palade (1952) has shown that with osmium fixatives the medium in which the fixative is dissolved precedes the fixative molecules into the tissue. The heavier molecules of osmium tetroxide lag behind. Since it is the osmium tetroxide which stabilizes (i.e. fixes) the cell constituents, these constituents are liable to damage from the medium if this arrives before the fixative molecules. It has, indeed, been empirically shown that with osmium fixatives the quality of fixation varies widely with the type of buffer used (Wood and Luft, 1965; Trump and Ericsson, 1965; Ericsson *et al.*, 1965). In most commonly used recipes for osmium fixatives, the particle concentration of the medium is increased to approximately

0·3 M by the addition of either sucrose (Caulfield, 1957) or glucose (Millonig, 1962).

Formaldehyde and glutaraldehyde, however, are light molecules. There is little reason to assume that the medium in which they are dissolved penetrates the tissue first, and the buffer employed makes less difference to the quality of fixation than with osmium fixatives. In a concentration of 4 per cent, which is frequently used for fixation, both of these aldehydes are hypertonic. Four per cent glutaraldehyde is 0·4 M; 4 per cent formaldehyde is 1·3 M, and the concentration of the medium must of course be added to this. Nevertheless, adequate fixation is obtainable using these concentrations. In fact, Sabatini *et al.* (1963) recommended addition of sucrose to formaldehyde fixatives for optimum preservation of ultrastructure.

Unexpected results are often obtained with hypertonic solutions. For instance, Burgos *et al.* (1967) during experiments with varying concentrations of aldehyde fixatives produced mitochondrial swelling with high concentrations rather than shrinkage. Immersion in 0·5 M urea causes cells to swell, even though its osmolarity exceeds that of blood plasma (Dick, 1966).

If a double fixation routine is employed, such as glutaraldehyde followed by osmium tetroxide, it is usually considered best practice to use the same buffer for both fixatives. This is not always done, however, and the pros and cons are difficult to assess. Whether or not to rinse off the primary fixative from the tissue before post-osmication is also a matter of debate. If the tissue is rinsed, the question arises whether the tonicity of the rinsing buffer should match that of the primary fixative (which is often hypertonic), or that of the osmium 'post-fixative', or whether some arbitrary value in between should be chosen. The prime consideration is, of course, to disturb the tissue as little as possible during processing. There appears to be no general agreement on the best way to manage this. Fahimi and Drochmanns (1965) report best results with moderately hypertonic rinsing solutions.

An example of how to adjust a buffer to a particle concentration of approximately 0·3 M is given in the Appendix, p. 200.

In summary, there is very little known about the effects which various fixative–buffer systems have upon intracellular molecular and ionic concentrations. The electron microscopist is for the moment in the position of having to decide for himself which fixation procedure gives the best results with the particular tissue he is studying. Any decision he makes is bound to be arbitrary at the present time.

For further reading see Maunsbach (1964, 1966*a* and *b*), Karlsson and Schultz (1965, 1966), Schultz and Karlsson (1965), Tahmisian (1964), Burgos *et al.* (1967), Elbers (1966) and especially the reviews by Trump and Ericsson (1965), Hopwood (1966), and Hayat (1970).

DEHYDRATION AND EMBEDDING

In order to cut the fixed tissue into slices thin enough for microscopy, it is necessary to infiltrate the tissue with an embedding material which will support it and minimize the damage caused by the cutting.

The embedding media most commonly used in electron microscopy are not miscible with water. Therefore if infiltration of the embedding medium into the tissue is to be accomplished, the water in the tissue must first be replaced by some material with which the embedding medium will mix. Various dehydrants, including acetone and methanol, have been used for this purpose, but it is now generally agreed that ethanol removes water from the tissues with less deleterious effect than the other dehydrants. However, since ethanol is a lipid solvent, dehydration must be accomplished as rapidly as possible to avoid the leaching of lipid-containing tissue components.

If the tissue were transferred direct from the fixative to 100 per cent ethanol, mechanical damage to the tissues might ensue due to the changes in surface tension and currents set up as the two fluids mix. Therefore, the tissue is passed through a graded series of ethanols, say 70, 90 and 100 per cent, to minimize the damage. This can be easily done by pipetting off or pouring off one fluid from the tissue replacing it with the next. Ten minutes in each percentage of ethanol, followed by a second and third bath of fifteen minutes each in 100 per cent ('absolute') ethanol is usually sufficient. Dehydration is facilitated if the vials containing the tissue and dehydrants are spun slowly on a turntable during this process.

If the embedding medium to be used is miscible with ethanol, direct transfer of the tissue between the two media can be made. If not, it will be necessary to pass the tissue into another fluid which is mutually miscible with ethanol and the embedding medium. Epoxypropane (propylene oxide) is used for this purpose when epoxy resins are used for embedding. A total of 5 minutes in epoxypropane with one change during this time is sufficient. The tissue is then transferred to a small plastic 'boat' containing unpolymerized embedding medium. This can be done without damage to the tissues if the tweezers used for transfer are first dipped into the embedding mixture so that a 'blob' of the material adheres to the end of the tweezers. The tissue can then be picked up in this 'blob' and transferred without the tweezers ever actually contacting the tissue.

The tissue must be left in unpolymerized embedding medium long enough for the medium to completely penetrate the tissue. A few hours or overnight at room temperature is generally adequate. When infiltration is complete, the tissue is transferred to fresh embedding medium, in whatever sort of containers the final polymerization is to occur. These are usually gelatin or polythene capsules, or special embedding boats or moulds. Capsules produce a block of plastic of convenient size and shape to fit the microtome chuck. The capsules or moulds are placed in an oven, usually at 60 °C, where polymerization (hardening) of the embedding medium will occur in the presence of the proper catalyst. Polymerization of some embedding media (i.e. the methacrylates) is more satisfactorily accomplished in a cold room using ultraviolet light as the polymerizing agent.

Types of Embedding Media

In order for the electron beam to penetrate a section of tissue, the section must be no thicker than 1000 Å (0·1 μm). Sections as thin as this cannot be obtained with the usual histological embedding media such as paraffin wax. Thus in the early days of electron microscopy a great deal of attention was paid to finding a suitable embedding material.

The Methacrylates

The first such material to come into common use for electron microscopy was butyl methacrylate (Newman, Borysko and Swerdlow, 1950). This is an acrylic plastic related chemically to Perspex. It was soon found that by combining butyl and methyl methacrylates in different proportions, embedding media could be produced with the proper degree of hardness or softness to enable various types of tissue to be sectioned to a thickness of 500 Å or less.

Much of the early exploratory work in electron microscopy was done with tissue embedded in the methacrylates. However, the methacrylates suffer from a number of disadvantages. Tissue shrinkage often approaches 20 per cent during polymerization. Polymerization tends to proceed unevenly, resulting in damage to the tissue. Unless special precautions are taken, bubbles of gas may collect around the tissue. Sections of methacrylate-embedded tissue are very liable to thermal drift (movement) when exposed to the electron beam. Sublimation of the methacrylate also occurs upon exposure to the beam. This increases specimen contrast, but it also contaminates the microscope. Moreover, methacrylate sections are too fragile to withstand the electron beam unless the supporting metal mesh grid upon which they are mounted is covered with a thin film of collodion or Formvar. This, of course, results in an increased thickness of material for the beam to penetrate and cuts down resolution.

The Epoxy Resins

In the mid 1950's Glauert *et al.* (1956, 1958) introduced Araldite as an embedding medium for electron microscopy, and Kushida (1959) and Finck (1960) introduced Epon. Use of these epoxy resins overcame many of the problems encountered with methacrylate embedding. Shrinkage and polymerization damage are greatly reduced, and gas bubbles are not a problem. The sublimation rate of the resins when exposed to the electron beam is low. Sections of epoxy-resin-embedded tissues are strong enough to be mounted on grids without a supporting membrane. Today Araldite or Epon are the routine embedding media in most laboratories. Various mixtures of the two are becoming increasingly popular (Mollenhauer, 1964; Anderson, 1965).

The Polyester Resins

These share the advantages of the epoxy resins in polymerizing uniformly with little shrinkage. Vestopal W (Ryter and Kellenberger, 1958) is the only one which is at all widely used. It is rather difficult to cut. However, it is recommended by Sjöstrand (1969) for high resolution work because its sublimation rate when exposed to the electron beam is very low. Thus distortion of tissue components caused by removal of embedding material is kept to a minimum.

For a further discussion of the non-water-soluble embedding media see Wachtel *et al.* (1966).

Water Soluble Embedding Media

If an embedding medium is water soluble it is not necessary to remove all the water from the tissues. The dehydration occurring during immersion of the tissue in increasing concentrations of the embedding medium is generally sufficient. Once embedded and sectioned, the tissue may be stained or subjected to cytochemical methods which cannot be used with non-water-soluble embedments. Of the water-soluble embedding media investigated—Aquon (Gibbons, 1959), Durcupan (Stäubli, 1960), glycol methacrylate (Leduc and Bernhard, 1962) and hydroxypropyl methacrylate (Leduc and Holt, 1965)—glycol methacrylate appears to be the most useful.

A more detailed account of the use of glycol methacrylate will be found in Chapter 8, and schedule for preparation and use on pp. 205–206 of the Appendix.

Effect of the Embedding Medium upon the Tissue

Since none of the embedding media are chemically inert, their possible effects on the tissues must be taken into account. A great deal has been written about the effects of the various fixatives upon tissues. Not so much has been heard about the effects of the various embedding media. This reflects the dearth of hard facts known about such effects, and the need for a great deal more research in this area. The methacrylates have a high sublimation rate when exposed to the electron beam and some materials from the tissue, particularly protein, come off in the sublimate. The use of epoxy and polyester resins reduces but does not eliminate this difficulty. The monomers of Epon, Araldite and the methacrylates have all been shown to be fat solvents. When dehydration of aldehyde-fixed tissue is carried out in increasing concentrations of water-soluble embedding media (as is often the case), glycogen is extracted (Leduc and Holt, 1965) and lipid components in the tissue disappear unless temperatures in the neighbourhood of −20°C are used during processing (Cope and Williams, 1968).

THE VALUE OF ADHERING TO A FIXED ROUTINE OF TISSUE PREPARATION

From the preceding discussion of preparative methods used in electron microscopy, it is clear that there are considerable differences of opinion about such matters as time and temperature of fixation, importance of pH and tonicity of the fixative, and the best routines for dehydration and embedding. It is also clear that there is an alarming number of variables to be considered from the time the animal to be used is selected until the ultrathin sections are viewed in the microscope. Ideally each tissue to be investigated should first be put through a series of carefully controlled experiments to determine exactly what effect each variable has on its ultrastructure. In practice this is seldom done except in laboratories where research is specifically on problems of this nature. Generally, a routine for tissue processing is adopted because it has worked well for some other person doing similar work.

The fact that there is no universal agreement on the one 'best' method of preparation does not mean that attention to detail and exact timings are not important. The best way of keeping the variables involved from undermining the value of your observations is to standardize the routine. Once a preparative routine has been decided upon for research involving a particular tissue it should be strictly adhered to. In this manner the variables involved are standardized and tissue processed at different times during the course of the particular research project can be meaningfully compared. Alterations of the routine should be made *only for the purpose of gaining further knowledge about the tissue*, not because it is nearly lunch time. When alterations are made, the details should be carefully recorded and the resulting micrographs scrutinized with the alterations in mind.

A TISSUE PROCESSING SCHEDULE RECOMMENDED FOR ROUTINE WORK

The following schedule for the routine processing of tissues is one which has been successfully used in our laboratory for a number of years. Detailed recipes for the fixatives, buffers, embedding media and stains employed will be found in the Appendix.

1. *Setting up for processing the tissue.* (a) Check all recipes to be used during processing, and assemble the required materials and glassware. Preparation of the fixatives, buffers, dehydrants and embedding media will have to be commenced a considerable time before removal of tissues from the animal. Make out a rough time table to be used as a guide. If fixation is to be done at 0°C, leave time for the fluid to be cooled to that temperature in an ice bath.

(b) Make up the necessary solutions and media.

(c) Assemble materials to be used during processing: dissecting instruments, razor blades for slicing tissue, a firm clean surface such as dental wax on which to slice, wide-mouth Pasteur pipettes for transferring tissues or removing fluids from tissues, small bottles (appropriately labelled) in which to fix and dehydrate, and embedding boats or capsules. The bottles should be scrupulously clean, especially if osmium-containing fixatives are to be used, since traces of organic matter on the glass will reduce osmium and render the fixative less effective.

(d) Turn on embedding oven to equilibrate to the required temperature (usually 60°C).

2. *Removing the tissue.* Anaesthetize the animal with veterinary Nembutal (0·1 ml per 100 g body weight injected intraperitoneally) and remove the tissue as described earlier.

3. *Fixation.* Two to 4 per cent buffered glutaraldehyde for 4 hours.

4. *Rinsing.* Remove the fixative with a Pasteur pipette and replace it with rinsing buffer. Alternatively, pour off the fixative carefully (into a beaker, not down the sink, in case pieces of tissue come out with the fluid!). For rinsing use the same buffer as used in the fixative, raising the particle concentration to approximately 0·3 M with sucrose or glucose. Rinse at least ½ hour with three changes of buffer.*

5. *'Post-fixation'.* One per cent osmium tetroxide (Millonig, 1962) for 1 hour.

6. *Rinsing.* Two quick changes of either Millonig's phosphate buffer or distilled water.

7. *Dehydration.* Ethyl alcohol (ethanol, absolute alcohol) as follows:

70 per cent—10 min
90 per cent—10 min
100 per cent—10 min
100 per cent—two changes, 15 min each.

(The 100 per cent (anhydrous) ethanol should be stored in contact with silica gel desiccant before use.)

Epoxypropane (propylene oxide)—5 min total, with one change. (Total dehydration time=65 min.)

8. *Infiltration.* Place the tissue in Araldite embedding mixture and leave overnight at room temperature.

9. *Embedding.* Move the tissue to fresh Araldite embedding mixture in capsules or embedding boats and place in an oven at 60°C for 24 hours. If capsules are used, one piece of tissue is placed in each capsule, and the capsules set upright in special holders so the tissue will sink to the bottom.

10. *Trimming.* Trim the polymerized blocks as directed in Chapter 3.

11. *Cutting and staining.* (a) One μm sections for orientation and study: Mount on glass microscope slides and stain with the quick methylene blue–borax method.

(b) Ultrathin sections for electron microscopy: Mount on metal mesh supporting grids and stain with aqueous uranyl acetate followed by Reynold's lead citrate stain.

The material is now ready to be viewed in the electron microscope.

* Sabatini *et al.* (1963) reported that tissues can be stored in rinsing buffer for several months at 4°C. However, this is not now recommended, since it has not been fully established what materials may leach out during such prolonged storage.

REFERENCES

Post mortem changes

Al Azzawi, H. T. & Stoward, P. J. (1970). The effect of various forms of euthanasia on the histochemically demonstrable activities of lysosomal enzymes. *Proceedings of the Microscopical Society* **5**, 13–14.

Ito, S. (1962). In *The Interpretation of Ultrastructure*, Vol. 1, pp. 129–148 (Edited by R. J. C. Harris), New York: Academic Press.

Karlsson, U. & Schultz, R. L. (1966). Fixation of the central nervous system for electron microscopy by aldehyde perfusion. III. Structural changes after exsanguination and delayed perfusion. *Journal of Ultrastructure Research* **14**, 47–63.

Malhotra, S. K. (1968). In *Cell Structure and its Interpretation*, pp. 381–391 (Edited by S. M. McGee-Russell and K. F. A. Ross), London: Edward Arnold.

Sjöstrand, F. S. & Hanzon, V. (1954). Membrane structure of cytoplasm and mitochondria in exocrine cells of mouse pancreas as revealed by high resolution electron microscopy. *Experimental Cell Research* **7**, 393–414.

Trump, B. F., Goldblatt, P. J. & Stowell, R. E. (1962). An electron microscopic study of early cytoplasmic alterations in hepatic parenchymal cells of the mouse liver during necrosis *in vitro* (autolysis). *Laboratory Investigation* **11**, 986–1015.

Trump, B. F., Goldblatt, P. J. & Stowell, R. E. (1965). Studies on necrosis of mouse liver *in vitro*. *Laboratory Investigation* **14**, 343–371.

Fixatives and fixation procedures (including osmolarity)

Afzelius, B. A. (1962). In *The Interpretation of Ultrastructure*, pp. 1–19. (Edited by R. J. C. Harris), New York: Academic Press.

Bahr, G. F. (1954). Osmium tetroxide and ruthenium tetroxide and their reactions with biologically important substances. Electron stains III. *Experimental Cell Research* **7**, 457–479.

Bahr, G. F. (1955). Continued studies about the fixation with osmium tetroxide. Electron Stains IV. *Experimental Cell Research* **9**, 277–285.

Bahr, G. F., Bloom, G. & Friberg, W. (1957). Volume changes of tissues in physiological fluids during fixation in osmium tetroxide or formaldehyde and during subsequent treatment. *Experimental Cell Research* **12**, 342–355.

Bennet, H. S. & Luft, H. (1959). *s*-collidine as a basis for buffering fixatives. *Journal of Biophysical and Biochemical Cytology* **6**, 113–114.

Bradbury, S. & Meek, G. A. (1960). A study of potassium permanganate 'fixation' for electron microscopy. *Quarterly Journal of Microscopical Science* **101**, 241–250.

Burgos, M. H., Vitale-Caple, R. & Tellezale-Inon, M. T. (1967). Studies on paraformaldehyde fixation for electron microscopy. 1. Effect of concentration on ultrastructure. *Journal de Microscopie* **6**, 457–468.

Burkl, W. & Schiechl, H. (1968). A study of osmium tetroxide fixation. *Journal of Histochemistry and Cytochemistry* **16**, 157–161.

Caulfield, J. B. (1957). Effects of varying the vehicle for osmium tetroxide in tissue fixation. *Journal of Biophysical and Biochemical Cytology* **3**, 827–820.

Dalton, A. J. (1955). A chrome–osmium fixative for electron microscopy. *Anatomical Record* **121**, 281 (abs.).

Dick, D. A. T. (1966). *Cell Water*, Chap. 3, pp. 15–43, London: Butterworth.

Elbers, P. F. (1966). Ion permeability of the egg of *Limnaea stagnalis* L. on fixation for electron microscopy. *Biochimica et biophysica Acta* **112**, 318–329.

Ericsson, J. L. E., Saladino, A. J. & Trump, B. F. (1965). Electron microscopic observation of the influence of different fixatives on the appearance of cellular ultrastructure. *Zeitschrift für Zellforschung un Mikroskopische Anatomie* **66**, 161–181.

Ericsson, J. L. E. & Biberfeld, P. (1967). Studies on aldehyde fixation. Fixation rates and their relation to fine structure and some histochemical reactions in liver. *Laboratory Investigation* **17**, 281–298.

Fahimi, H. D. & Drochmanns, P. (1965). Essais de standardisation de la fixation au glutaraldehyde. II. Influence des concentrations en aldehyde et de l'osmolarité. *Journal de Microscopie* **4**, 737–748.

Fawcett, D. W. (1964). In *Modern Developments in Electron Microscopy*, Chap. 6, p. 309 (Edited by B. M. Siegel), New York and London: Academic Press.

Friend, D. S. & Farquhar, M. G. (1967). Function of coated vesicles during protein absorption in the rat vas deferens. *Journal of Cell Biology* **35**, 357–376.

Hagström, L. & Bahr, G. F. (1960). Penetration rates of osmium tetroxide with different fixation vehicles. *Histochemie* **2**, 1–4.

Hake, T. (1965). Studies on the reactions of osmium tetroxide and potassium permanganate with amino acids, peptides, and proteins. *Laboratory Investigation* **14**, 1208–1212.

Hayat, M. A. (1970). *Principles and Techniques of Electron Microscopy: Biological Applications*, Vol. 1. New York and London: Van Nostrand-Reinhold.

Hopwood, D. (1967). Some aspects of fixation with glutaraldehyde. *Journal of Anatomy* **101**, 83–92.

Hopwood, D. (1969). Fixatives and fixation: a review. *Journal of Histochemistry and Cytochemistry* **1**, 323–360.

Karlsson, U. & Schultz, R. L. (1965). Fixation of the central nervous system for electron microscopy by aldehyde perfusion. I. Preservation with aldehyde perfusates versus direct perfusion with osmium tetroxide, with special reference to membranes and the extra-cellular space. *Journal of Ultrastructure Research* **12**, 160–186.

Karnovsky, M. J. (1965). A formaldehyde-gultaraldehyde fixative of high osmolarity for use in electron-microscopy. *Journal of Cell Biology* **27**, 137A–138A.

Korn, E. D. (1967). A chromatographic and spectrophometric study of the products of the reaction of osmium tetroxide with unsaturated lipids. *Journal of Cell Biology* **34**, 627–638.

Luft, J. H. (1956). Permanganate—a new fixative for electron microscopy. *Journal of Biophysical and Biochemical Cytology* **2**, 799–801.

Luft, J. H. & Wood, R. L. (1963). The extraction of tissue protein during and after fixation with osmium tetroxide in various buffer systems. *Journal of Cell Biology* **19**, 46A.

Maunsbach, A. B. (1964). Comparison of renal tubule ultrastructure after perfusion fixation with different fixatives. *Journal of Cell Biology* **23**, 108A–109A.

Maunsbach, A. B. (1966*a*). The influence of different fixatives and fixation methods on the ultrastructure of rat kidney proximal tubule cells. I. Comparison of glutaraldehyde, formaldehyde and osmium tetroxide fixatives. *Journal of Ultrastructure Research* **15**, 242–282.

Maunsbach, A. B. (1966*b*). The influence of different fixatives and fixation methods on the ultrastructure of rat kidney proximal tubule cells. II. The effects of varying osmolarity, ionic strength, buffer systems and fixative concentrations of glutaraldehyde solutions. *Journal of Ultrastructure Research* **15**, 283–309.

Millonig, G. (1962). In *Fifth International Congress in Electron Miscroscopy*, Vol. 2, p. 8. (Edited by S. S. Breese), New York: Academic Press.

Palade, G. E. (1952). A study of fixation for electron microscopy. *Journal of Experimental Medicine* **95**, 285–298.

Palay, S. L., McGee-Russell, S. M., Gordon, S. J. & Grillo, M. A. (1962). Fixation of neural tissues for electron microscopy by perfusion with solutions of osmium tetroxide. *Journal of Cell Biology* **12**, 385–410.

Pearse, A. G. E. (1968). In *Histochemistry, Theoretical and Applied*, 3rd edn, Chap. 5, London: J. and A. Churchill.

Pease, D. C. (1962). Buffered formaldehyde as a killing agent and primary fixative for electron microscopy. *Anatomical Record* **142**, 342 (abs.).

Porter, K. R. & Kallman, T. (1953). The properties and effects of osmium tetroxide as a tissue fixative with special reference to its use for electron microscopy. *Experimental Cell Research* **4**, 127.

Richards, F. M. & Knowles, J. R. (1968). Glutaraldehyde as a protein cross-linking reagent. *Journal of Molecular Biology* **37**, 231–233.

Riemersma, J. C. (1968). Osmium tetroxide fixation of lipids for electron microscopy: a possible reaction mechanism. *Biochimica et biophysica Acta* **152**, 718–727.

Roth, L. E. *et al.* (1963). Glutaraldehyde–osmium tetroxide fixation for electron microscopy of the mitotic apparatus in plant and insect cells. *Journal of Cell Biology* **19**, 91A.

Sabatini, D. D., Bensch, K. & Barnett, R. J. (1963). Cytochemistry and electron microscopy—the preservation of cellular ultrastructure and enzymatic activity by aldehyde fixation. *Journal of Cell Biology* **17**, 19–58.

Sabatini, D. D., Miller, F. & Barnett, R. J. (1964). Aldehyde fixation for morphological preservation and enzyme histochemical studies with the electron microscope. *Journal of Histochemistry and Cytochemistry* **12**, 57–71.

Schultz, R. L. & Karlsson, U. (1965). Fixation of the central nervous system for electron microscopy by aldehyde perfusion. II. Effect of osmolarity, pH of perfusate, and fixative concentration. *Journal of Ultrastructure Research* **12**, 187–206.

Schultz, R. L. & Case, N. M. (1970). A modified aldehyde perfusion technique for preventing certain artefacts in electron microscopy of the central nervous system. *Journal of Microscopy* **92**, 69–84.

Sjöstrand, F. S. (1962). In *The Interpretation of Ultrastructure*, pp. 47–68 (Edited by R. J. C. Harris), New York and London: Academic Press.

Stoeckenius, W. & Mahr, S. C. (1965). Studies on the reaction of osmium tetroxide with lipids and related compounds. *Laboratory Investigation* **14**, 1196–1207.

Tahmisian, T. N. (1964). Use of the freezing point method to adjust the tonicity of freezing point solutions. *Journal of Ultrastructure Research* **10**, 182–188.

Tandler, C. J., Libanati, C. M. & Sanchis, C. A. (1970). The intracellular localisation of inorganic cations with potassium pyroantimonate. *Journal of Cell Biology* **45**, 355–366.

Trump, B. F. & Ericsson, J. L. E. (1965). The effect of the fixative solution on the ultrastructure of cells and tissues. *Laboratory Investigation* **14**, 1245–1323.

Trump, B. F. & Bulger, R. E. (1966). New ultrastructural characteristics of cells fixed in a glutaraldehyde–osmium tetroxide mixture. *Laboratory Investigation* **15**, 368–379.

Weibel, E. R. (1969). Stereological principles for morphology in electron microscopic cytology. *International Review of Cytology* **26**, 235–302.

Williams, M. A. (1969). In *Advances in Optical and Electron Microscopy*, Vol. 3, pp. 219–228 (Edited by R. Barer and V. E. Cosslett), London and New York: Academic Press.

Wood, R. L. & Luft, J. H. (1965). The influence of buffer systems on fixation with osmium tetroxide. *Journal of Ultrastructure Research* **12**, 22–45.

Embedding media

Cope, G. H. & Williams, M. A. (1968). Quantitative studies on neutral lipid preservation in electron microscopy. *Journal of the Royal Microscopical Society* **88**, 259–277.

Finck, H. (1960). Epoxy resins in electron microscopy. *Journal of Biophysical and Biochemical Cytology* **7**, 27–30.

Gibbons, I. R. (1959). An embedding resin miscible with water for electron microscopy. *Nature, London* **184**, 375–376.

Glauert, A. M., Rogers, G. E. & Glauert, R. H. (1956). A new embedding medium for electron microscopy. *Nature, London* **178**, 803.

Glauert, A. M. & Glauert, R. H. (1958). Araldite as an embedding medium for electron microscopy. *Journal of Biophysical and Biochemical Cytology* **4**, 191–194.

Leduc, E. H. & Bernhard, W. (1962). In *The Interpretation of Ultrastructure*, pp. 21–45 (Edited by R. J. C. Harris), New York and London: Academic Press.

Leduc, E. H. & Holt, S. J. (1965). Hydroxypropyl methacrylate, a new water miscible embedding medium for electron microscopy. *Journal of Cell Biology* **26**, 137–155.

Luft, J. H. (1961). Improvements in epoxy resin embedding methods. *Journal of Biophysical and Biochemical Cytology* **9**, 409–414.

Kushida, H. (1959). On an epoxy resin embedding method for ultrathin sectioning. *Electron Microscopy* **8**, 72–74.

Mollenhauer, H. H. (1964). Plastic embedding mixtures for use in electron microscopy. *Stain Technology* **39**, 111-114.

Newman, S. B., Borysko, E. & Swerdlow, M. (1950). Ultramicrotomy by a new method. *Journal of Applied Physiology* **12**, 67 (abs.).

Ryter, A. & Kellenberger, E. (1958). L'inclusion au polyester pour l'ultramicrotomie. *Journal of Ultrastructure Research* **2**, 200–214.

Stäubli, W. (1960). Nouvelle matière d'inclusion hydrosoluble pour la cytologie électronique. *Comptes rendu hebdomadaire des séances de l'Académie des sciences* **250**, 1137–1139.

Wachtel, A. W., Gettner, M. E. & Ornstein, L. (1969). In *Physical Techniques in Biological Research*, 2nd edn, Vol. 3, Part A Cells and Tissues, Chap. 4 (Edited by A. W. Pollister), New York and London: Academic Press.

For further references see bibliography at end of Chapter 8.

General discussions of tissue processing

Bahr, G. F. & Zeitler, E. H. (Eds.) (1965). *Proceedings of a Symposium on Quantitative Electron Microscopy.* Seven papers on the topic 'Effects of the preparation procedure on the appearance of the object in electron micrographs'. *Laboratory Investigation* **14**, 1169–1245.

Glauert, A. M. (1965). In *Techniques for Electron Microscopy*, Chap. 7, pp. 166–212 (Edited by D. H. Kay), Oxford: Blackwell Scientific Publications.

McGee-Russell, S. M. & De Bruijn, W. C. (1968). In *Cell Structure and its Interpretation* (Edited by S. M. McGee-Russell and K. F. A. Ross), London: Edward Arnold.

Millonig, G. & Marinozzi, V. (1968). In *Advances in Optical and Electron Microscopy*, Vol. 2, pp. 251–341 (Edited by R. Barer and V. E. Cosslett), London and New York: Academic Press.

Pease, D. C. (1964). In *Histological Techniques for Electron Microscopy*, 2nd edn, Chap. 2, 3 and 4, New York and London: Academic Press.

Sjöstrand, F. S. (1967). In *Electron Microscopy of Cells and Tissues*, Vol. 1, Chap. 6 and 7, pp. 138–176, New York and London: Academic Press.

Sjöstrand, F. S. (1969). In *Physical Techniques in Biological Research*, Vol. 3, Part C, Cells and Tissues, pp. 169–189 (Edited by A. W. Pollister), New York and London: Academic Press.

Wischnitzer, S. (1967). Current techniques in biomedical electron microscopy. *International Review of Cytology* **22**, 1–61.

See also references at end of Chapter 8.

3

The Preparation of Ultrathin Sections for Electron Microscopy

In many ways the most difficult and frustrating part of electron microscopy is the sectioning of tissues. It is a matter of opinion whether prior experience in sectioning material for light microscopy is helpful or not. Most of the manoeuvres during the preparation of ultrathin sections will be performed with the aid of a stereomicroscope, so one must get used to working with this essential piece of equipment. The most important single prerequisite for success is patience. When patience is exhausted, it is wisest to leave the microtome for a while, do something else, and come back fresh.

The time required to become proficient at sectioning varies widely. Some talented and productive electron microscopists have never mastered the art, and leave their sectioning to well-trained technicians. The collection of traits necessary for successful sectioning does not necessarily overlap that required for the interpretation of ultrastructure. Nevertheless, if an electron microscopist is an expert at the ultramicrotome he will be far better able to train his technical assistants to a high standard, and to cut his own especially critical experimental tissues should the need arise. Several persons of outstanding reputation in the field insist upon cutting their own sections.

The average beginner usually manages to get some passable sections within two weeks. A few succeed in their first attempt. Only rarely has our staff had to do a student's sectioning for him to enable him to complete his project.

THE NEED FOR ORIENTATION SECTIONS

In order to be sure that you have found a portion of the specimen which contains the material you wish to study, you must take preliminary 'thick' sections and look at them with a light microscope. Fortunately, the qualities of the acrylic plastics and epoxy resins used for embedding the tissue are such that 1 μm sections can be taken from the specimen with a glass knife. After appropriate staining, the detail seen in these sections is far greater than that seen in paraffin sections of similar tissue. By referring to the 1 μm section one can trim the specimen so that ultrathin sections will include only pertinent material. The 1 μm sections can be cut either on a conventional microtome modified to take a glass knife, or on the ultramicrotome itself. In our own laboratory, we have found it most convenient for routine work to do the two types of sectioning on different machines, thus alleviating the pressure of work on the ultramicrotomes. If sections less than 1 μm thick are desired these are cut on the ultramicrotome, since conventional microtomes are seldom constructed to produce sections less than 1 μm thick. At any rate, sections thinner than 1 μm have so little tissue in them that staining for light microscopy becomes progressively less satisfactory as thickness decreases.

GLASS KNIVES

Glass knives were introduced by Latta and Hartmann (1950). They are far superior to metal knives for cutting ultrathin sections of tissues embedded in acrylic plastics or epoxy resins. They can be used for a short time only before the edge deteriorates, but are sufficiently inexpensive to be discarded after use. These knives must also be used within a short time after making, since glass is a supercooled liquid which will gradually flow and the edge will become blunt.

Methods of Making Glass Knives

Knives for both 1 μm sections and ultrathin sections for electron microscopy (100 to 1200 Å thick) are made in the same way from $\frac{1}{4}$ inch or $\frac{3}{16}$ inch plate glass or float glass. The glass is carefully washed in detergent, rinsed in hot water and allowed to dry in air. This will remove grease, dust and other material which might otherwise contaminate the tissue sections during cutting. The glass should not be wiped dry, as this will set up an electrostatic charge which will attract dust.

The glass is broken into squares or rectangles of the appropriate size. From these pieces, the glass knives are made.

Knife-making Machines

Knives can be made either by hand, or by using a special knife-making machine such as that produced by LKB Instruments Ltd. A knife-making machine is simple to use once it has been properly set up, and yields a very high percentage of excellent knives. One only need follow the directions in the manual accompanying the knife-making machine. The manufacturer supplies strips of specially selected high-quality glass to be used with it. A knife-making machine is expensive and so is the glass provided for use with it. However, it saves time and frustration in the long run in a laboratory where many people are making knives.

Breaking Knives by Hand

If a knife-making machine is not available, the knives must be broken by hand. This takes practice and the failure rate will be high at first.

Getting a supply of the right quality glass can be a problem. Glass made by either the plate or 'float' method will vary from one batch to another. If you find a sample of glass which cuts well, try to lay in a supply of the same batch. Knives can appear to be perfect by all the usual criteria (see p. 50) and still not cut good sections if the glass is from a poor batch. Unfortunately the only completely reliable test for a good knife is its production of good sections.

Knives for cutting 1 μm sections and for cutting ultrathin sections are made by the same methods. The dimensions of the pieces used for the final break will vary with the type of microtome used. Most modern ultramicrotomes require 1 inch squares; some machines, such as the older models of the Cambridge Huxley manual ultramicrotome, require rectangles measuring 1 inch by $1\frac{1}{2}$ inches.

For breaking glass you will need a wheel-type glass-cutter and a pair of specially adapted glass-breaking pliers. The wheels of the glass-cutter should be in excellent condition. Glass-breaking pliers specially machined for production of glass knives are made by both Sorvall and Cambridge Instruments (see Appendix, p. 222). Alternatively, ordinary glaziers' pliers may be adapted for knife-making by using tape, as described by Porter (1964). The taping must be redone periodically, however, whereas the specially machined pliers require no maintenance. However, taping allows adjustment of the pliers to the individual's handgrip strength, and a slow, controllable break may be assured by using the proper amount and distribution of tape.

A guide with a straight edge is needed for use with the glass-cutter in order to make straight scoring lines. A metal ruler will suffice, or guides may be made from Perspex (Lucite) to any desired dimensions. If the knife-making method being used calls for strips of glass 2 inches wide, such strips can be produced by laying a straight-edged guide $1\frac{7}{8}$ inches wide by, say, 8 inches long on a sheet of glass with the long dimension of the guide coinciding with the edge of the glass. A score is made along the guide with the glass-cutter, and the glass is then broken with the pliers. Then the process is repeated until the sheet has been broken into 2-inch strips. (The glass-cutter usually requires $\frac{1}{8}$ inch clearance, so the straight-edged guide is made $\frac{1}{8}$ inch narrower than the width of the strip desired. For example, if a strip $1\frac{1}{2}$ inches wide is required the guide should be $1\frac{3}{8}$ inches wide. If a 1 inch strip or 1 inch squares are to be broken the guide should be $\frac{7}{8}$ inch wide.) A small guide measuring $\frac{7}{8}$ inch by, say, 5 inches is convenient for breaking small pieces of glass. When laid consecutively along a strip of glass 1 inch, $1\frac{1}{2}$ inches or 2 inches wide for purposes of scoring, this guide will produce 1 inch squares, $1\frac{1}{2} \times 1$ inch rectangles, or 1×2 inch rectangles depending on the width of the strip.

There are a number of methods of breaking the glass into the final shape. Two of these will be briefly described in this chapter.

1. *The 'Free Break' Method*

This method was introduced by Cameron (1956). It produces (in expert hands) knives of superior quality. There are many variations of this method, and the one described here is basically that advocated by Porter (1964):

A square piece of glass measuring 4 inches on a side is broken into halves and then into halves again until 16 pieces 1 inch square are obtained. A final diagonal break produces a knife from each square. To facilitate breaking, short straight scores with the glass-cutter are made at the periphery of the glass. For the first break, a straight score-mark $\frac{1}{2}$ inch long is produced with a straight-edge and glass-cutter at right angles to the edge, half way along one side of the 4 inch square. An even, slowly increasing pressure is then exerted on the score-mark with the glass-breaking pliers (Fig 7*a*). Be sure that the front edge of the pliers is perpendicular to the score-mark to assure a straight break. Slowly increasing, even, pressure will cause a slow, controllable break. The glass should break straight across, bisecting the glass.

Since the break did not follow a score-mark all the way across the glass, it is known as a 'free break' (Fig. 7*b*). The edges of the break will be much smoother than one following a score-mark. Had the score extended all the way across the glass, the break would have occurred along the scored line and the edge of the break would exhibit tiny fracture marks ('feathering') produced by the glass-cutter.

The following breaks are made in the same manner until 1 inch squares are produced. When making each break, make sure to keep the free-break side of the glass opposite the pliers (Fig. 7*c*). If this is done, all of the 1 inch squares finally produced by the 'free break' method will have at least two smooth sides meeting at right angles with no score feathering on either side of the angle. A knife produced by breaking diagonally into such a 'clean corner' will have a cutting edge which is usually superior to one produced from a diagonal break into a 'feathered corner'. This is because stresses set up within the glass by the scoring can induce irregularities along the knife edge which is finally produced.*

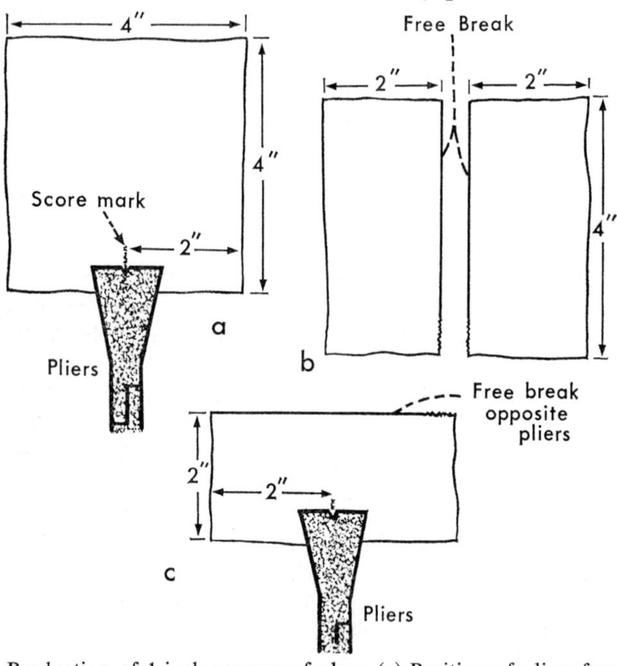

Fig. 7. Production of 1-inch squares of glass. (*a*) Position of pliers for initial break. (*b*) The resulting 'free break'. (*c*) Position of pliers for subsequent breaks.

* The strips of glass provided with most knife-making machines have feathered edges: nevertheless, as long as the knives are made according to the manufacturer's instructions and the knife-making machine is in proper adjustment quite satisfactory knives are produced. The important variable here appears to be the depth of the score-marks. Heavy pressure on a glass-cutter will produce a deep score and set up considerable stress within the glass. A light, even pressure will produce less stress (Sjöstrand, 1967). The manufacturers of knife-making machines have, of course, taken this into consideration in producing their strips of glass and in the design of their machines. The proper amount of pressure on a scoring tool is much easier to attain with a properly adjusted machine than by hand.

Producing 1 inch squares is relatively easy. Much more difficult is obtaining a good knife by breaking the 1 inch squares diagonally. Very slight deviations of the score-mark from the true diagonal will affect the shape of the knife edge. If the pliers are not pointed directly along the score-mark and centred over it, this too will affect the edge produced.

The diagonal break is made by scoring along a straight-edged guide, starting within about a millimetre of a 'clean corner' (Fig. 8*a*) and then breaking with the pliers. The front edge of the pliers

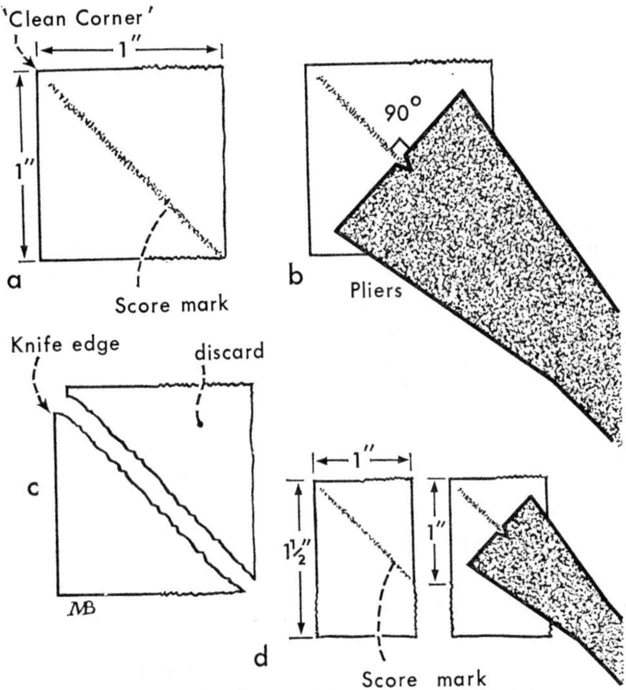

Fig. 8. Production of a knife from a 1-inch square. (*a*) Position of the score mark (*b*) Position of pliers. (*c*) Knife produced by break, side view. (*d*) Position of final diagonal scoring line if a rectangle is used.

should be positioned approximately half way along the score-mark (Fig. 8*b*) and a slow, steadily increasing pressure exerted. This final break results in one piece of glass bearing a sharp cutting edge and a second piece which is discarded (Fig. 8*c*).

If rectangles are desired instead of squares, the initial dimensions of the glass should be altered to 4 inches by 6 inches and four free-broken strips $1\frac{1}{2}$ inches wide produced. If score-marks at 1 inch intervals are made along these strips, four $1\frac{1}{2}$ inch by 1 inch rectangles will be produced from each strip. The final diagonal scoring line on the rectangle is positioned as in Fig 8*d*.

If this version of the free break method is to be used, reading the detailed discussion by Porter (1964) is recommended. Another variant is described in detail by Sjöstrand (1967).

2. *Pyper's Adaptation of the Free Break Method*

Obtaining a first-class knife by breaking a 1 inch square diagonally after scoring with a straight edged guide and glass-cutter is difficult and requires practice and patience. Getting a good knife is much simpler if a jig is used to position the glass for the final scoring, thus assuring a straight score on the true diagonal. An adaptation of the free break method by Pyper (1970) employs such a jig.

Strips of glass 2 inches wide are produced by scoring a sheet of glass along a straight edge and breaking with pliers. The 2 inch wide strip is then broken into 2 inch squares, also by scoring along a straight edge. Successive free breaks within each 2 inch square result in the production of four 1 inch squares, all with at least one clean corner. The knife is produced from a 1 inch square by use of a special jig which controls the position of the final score-mark in relation to the clean corner.

The method is quick, simple and produces a large percentage of good knives. Pyper describes production of the jig and pliers in sufficient detail that they may be made in a departmental workshop. Alternatively, Pyper's knifebreaking kit can be obtained from Cambridge Scientific Instruments Ltd (see Appendix, p. 222) who sell it as an 'optional extra' for the Cambridge Huxley Mk II ultramicrotome. Pyper's adaptation of the free break method may prove to be the method of choice for those laboratories which do not have a knife-making machine.

Assessing the Qualities of a Glass Knife

Without a high-quality cutting edge on the glass knife, good sections cannot be obtained. *Do not waste time trying to cut tissues with a knife that is at all questionable.*

Knife edges should always be checked under a stereomicroscope. A magnification of 20 times is usually adequate for routine work. Higher magnifications (up to 400 times) are necessary if the thinnest possible sections for high resolution microscopy are desired (Wachtel *et al.*, 1966). The cutting edge of the knife is brought into focus and the knife rotated slowly backward and forward until light is reflected off the edge. (Or the knife may be held stationary and the illumination adjusted.) When the light is reflecting off the edge, imperfections in the edge will be clearly visible.

An ideal knife edge should be very thin, straight across, and should have no nicks or irregularities. There should be no fracture lines running up the back or front of the knife to meet the edge. On the front face of the knife (i.e. the surface produced by the final diagonal break) there will be one stress line following a curved path which runs into one corner of the cutting edge (Fig. 9).

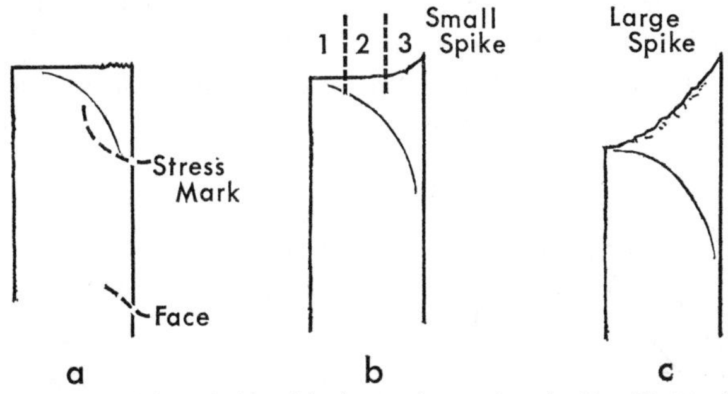

Fig. 9. Assessing a glass knife. (*a*) A nearly perfect knife. (*b*) Usable knife with small spike. 1=portion to be used for sectioning, 2=portion to be used for preliminary cuts, 3=sawtooth region. (*c*) Unusable knife with large spike.

The portion of the knife edge which will yield the best sections extends out from this corner for a variable distance. Near the corner opposite to that which intersects the stress line the edge will have a 'sawtooth' appearance and be useless for cutting. A short spike of glass may be present at this corner. This is caused by a slight displacement of the final score line from the true diagonal of the 1 inch square (or $1\frac{1}{2}$ inch rectangle). Ideally, a spike should not be present, since the length of the sawtooth region increases with the length of the spike. A small spike does not preclude the use of a knife. A knife bearing a large spike will have a very short usable cutting edge of questionable quality.

If a small spike is present, one can usually divide the knife edge roughly into thirds according to the quality of the cutting edge. The third nearest the spike is not used at all. The middle third is best used for preliminary cuts from the block. Once sections are coming off well, the knife may be shifted to the third nearest the stress mark corner so that the best part of the edge is used.

Since knives can only be used for a short time before the cutting edge deteriorates, it is good practice to break two or three good knives at one sitting. Then a knife can be replaced with minimum loss of time during sectioning. If a knife is not used on the day it is broken it should not be employed for ultrathin sectioning, since flowing of the glass slowly decreases the sharpness of the cutting edge. It may, however, be used for cutting 1 μm sections.

Once made, the knives must be handled with great care in order not to damage the cutting edge. They should be placed upright into special covered knife holders, or carefully placed on their sides on filter paper in a covered petri dish. Before they are used in the microtome, a 'boat' for holding the liquid onto which the sections are floated must be attached to the face of the knife below the cutting edge. For cutting 1 μm sections, a boat consisting of a film of molten dental wax, beeswax or paraffin wax is sufficient (Fig. 10*a*). For ultrathin sections a more satisfactory boat may be

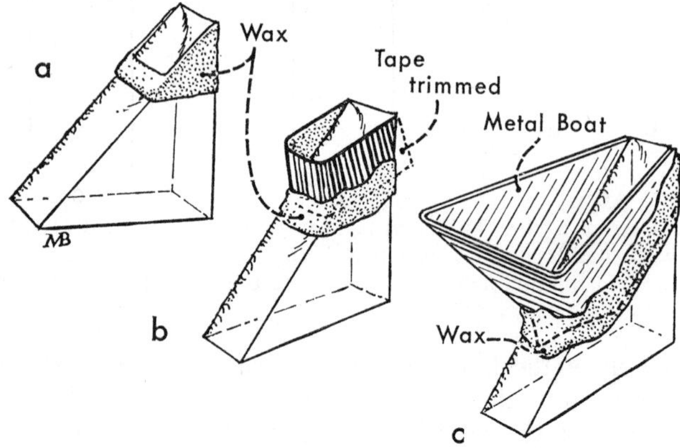

Fig. 10. Boats for glass knives. (*a*) Boat for 1 μm sections: a smear of wax. (*b*) Tape boat for ultrathin sections. (*c*) Metal boat for ultrathin sections.

constructed from a piece of black vinyl plastic electrical tape, e.g. Scotch Brand no. 33, fitted onto the knife face as shown in Fig. 10*b* and secured with molten wax.

Wax for securing the boat can be melted over a bunsen flame, but care should be taken that it does not get too hot. Not only is this a fire hazard, but the excess heat changes the properties of some waxes so that a greasy material may dissolve out of it onto the liquid in the boat and contaminate the sections. To avoid overheating, remove the wax from the flame before all of it has melted. A metal applicator (e.g. a knife with a non-heat-conducting handle) heated in the flame is convenient for putting the wax on the glass knife. Care must, of course, be taken not to get the wax on the knife edge.

Metal boats are supplied by the manufacturers of some ultramicrotomes (Fig. 10*c*). These also may be secured to the knife face with molten wax. Before re-use, these metal boats must be carefully cleaned. Most of the wax can be gently scraped off and the remainder removed with xylene. The xylene is then removed with ethanol. The knives may be stored in clean ethanol until needed.

DIAMOND KNIVES

The use of diamond knives was pioneered by Fernandez-Moran (1953). The cutting edge of such a knife has, of course, a far longer life than one made of glass and is more suitable for sectioning very hard tissues. The knives can be resharpened and may have a useful life of several years. However, the quality of the diamond knives commercially obtainable is highly variable. This fact and their high cost has kept them from being universally used.

SECTIONING

Attention to cleanliness is essential at all stages in the preparation of ultrathin sections. The ultrathin sections which are the final result of these procedures will be studied at magnifications where objects measuring less than 100 Å will be discernible. Any foreign material which gets on the sections can render hours of careful work useless. It is good practice to wipe working surfaces free from dust with a damp sponge before commencing work. Make sure instruments and pipettes are clean, and that tweezer points are in perfect condition. Fresh acetone solutions should be made up daily and kept stoppered to avoid contamination when not being used. It is safest to use Analar acetone, as other grades may contain traces of contaminating material.

If contamination is traced to dirty grids (some batches arrive from the manufacturers in less than ideal condition) see instructions for cleaning grids, Appendix, p. 217.

Materials Needed During Sectioning

A. *Materials for Both 1 μm and Ultrathin Sectioning*

1. A fresh 10 per cent solution of acetone in glass-distilled water. (Some workers recommend 20 per cent acetone or 10–20 per cent ethanol; others prefer distilled water. The addition of acetone or alcohol reduces the surface tension of water and makes the sections easier to manipulate.)

2. A Pasteur pipette or a syringe with a long needle for adding acetone solution to the boat.

3. A second pipette or syringe for removing acetone solution from the boat.
4. A beaker for discarding acetone solution removed from the boat.

B. *Materials for 1 μm Sectioning Only*

1. Clean glass microscope slides.
2. A holder (e.g. pliers or forceps) for holding the microscope slides in the bunsen flame.
3. Chemicals for staining the 1 μm sections.

C. *Materials for Ultramicrotomy Only*

1. A small bottle of chloroform with a glass rod attached to the stopper for use as an applicator. (A little cotton wool on the end of an orangewood stick will also suffice as an applicator.) Chloroform vapour is used for flattening sections after cutting.
2. An eyelash secured to the end of an orangewood stick with wax. This is excellent for manipulating sections on the acetone surface in the boat.
3. A pair of fine tweezers, in perfect condition, for handling grids.
4. A clean, dust-free dish lined with filter paper for holding specimen support grids.
5. Specimen support grids, either bare or coated with a collodion or Formvar film depending upon the type of grid and the nature of the embedding material. For routine work with material embedded in epoxy resin such as Araldite or Epon, coating the grids is unnecessary as long as fairly fine-mesh grids are used. Sections of tissue embedded in the methacrylates, however, are unstable in the electron beam unless extra support is given. Special techniques such as autoradiography (see Chapter 9) also require coated grids. The film must, of course, be as thin as possible to allow passage of the electron beam through both the film and the specimen. Methods of coating grids are given in the Appendix, pp. 212–215.
6. A second petri dish lined with filter paper for holding the grids after sections have been applied to them.
7. Container(s) for storage of grids after sectioning has been completed (i.e. gelatin capsules or a special grid box with slots for each grid). Gelatin capsules have been said to cause a film to form on the specimen surface after long storage, but this does not appear to be a hazard for storage periods less than six months.

Cutting 1 μm-Thick Sections

Whether these are cut on a conventional microtome or on the ultramicrotome the process is the same, although the controls to be manipulated will differ from one machine to another. The specimen advance on the microtome should be set at 1 μm. If the specimen has been embedded in capsules, the result will be a solid cylindrical block of material with the tissue positioned at one end. The capsule is removed (stripped off after slitting one side with a razor blade if the capsule is made of polythene; or melted off with warm water if the capsule is made of gelatin). The block is placed in a special holder, tissue facing outward, and the holder is fastened securely to the microtome. The knife is now mounted in its holder on the microtome. The back of the knife faces the specimen; the front (diagonal) face bearing the boat faces the operator. The back of the knife inclines toward the specimen at an angle not exceeding 6–7° from the vertical. The position of the tissue in relation to the knife is checked under a stereomicroscope mounted in front of the microtome. The block face containing the tissue should be parallel to the knife edge.

The knife is brought as close to the block as possible without actually touching it, and secured in position. The block is then advanced toward the knife edge in 1 μm increments while being raised and lowered with respect to the knife edge until a section is cut from the block face.

Now a drop of 10 per cent acetone is placed on the face of the knife above the wax boat. This will provide the meniscus on which the sections are floated. The shape of the meniscus is critical. If it is too convex, acetone solution will be dragged down the back of the knife as the block descends, taking the sections along with it. If the meniscus is too concave acetone will not quite reach the knife edge and the sections will adhere to the knife face in front of the edge.

After four or five sections have been cut they may be removed from the meniscus with a fine brush and transferred to a drop of acetone solution on a clean glass microscope slide. The slide is then passed rapidly through the flame of a bunsen burner to evaporate the acetone solution from beneath the sections. Ten or fifteen passages should be sufficient. Do not allow the slide to become too hot or the acetone will boil off, leaving bubbles under the sections. The sections once dried flat on the slide adhere firmly to the glass and can be treated with various solutions without fear of displacing them.

An alternative method is to coat a slide with a thin film of albumen, place the sections in a drop of water on the slide, and let them dry on a slide-warming tray. This sometimes results in fewer wrinkled sections than does the bunsen burner method, but it is slower.

Staining 1 μm-Thick Sections

After they are dry, the sections can be stained for study with a light microscope. A simple and suitable staining technique for this purpose is as follows:

Place equal numbers of drops of 1 per cent methylene blue and 1 per cent borax on the section. Pass the slide quickly through the flame of a bunsen burner about a dozen times. *Do not let the stain boil.* Rinse with a fine gently-running stream of hot tap water, and dry the underside of the slide with a clean cloth or tissue.

This preparation can be viewed at once with a light microscope and an appropriate area of the section selected. The block face can then be trimmed to include only this area. In this way one can ensure that the tissue which is finally incorporated into ultrathin sections contains pertinent material.

Other methods of staining 1 μm sections are given in the Appendix.

Trimming the Block for Ultrathin Sectioning

With a little practice it is easy to trim a block to a suitable area. A stereomicroscope capable of magnifications of twenty times or more, and a conventional light microscope are placed side by side. The 1 μm section is located under the light microscope with either the 10X or 40X objective. The old-fashioned type of microscope stage bearing clips, or a rotating stage, is an advantage, since the orientation of the slide can be changed freely. The block of embedded tissue is clamped, with its cut surface uppermost, in a holder to keep it steady. It is then placed under the low-power stereomicroscope and the cut face of the block is illuminated. By slowly turning the block about its vertical axis a point will be reached where the light is reflected from the surface at such an angle that details in the fixed tissue will stand out with considerable clarity. The orientation slide can then be turned until the section is aligned in the same direction as the block (Fig. 11). (If the section has been turned over during transfer to the slide, one will, of course, get a mirror image of what is seen on the block face.)

The block face is easily reduced to the desired area with a

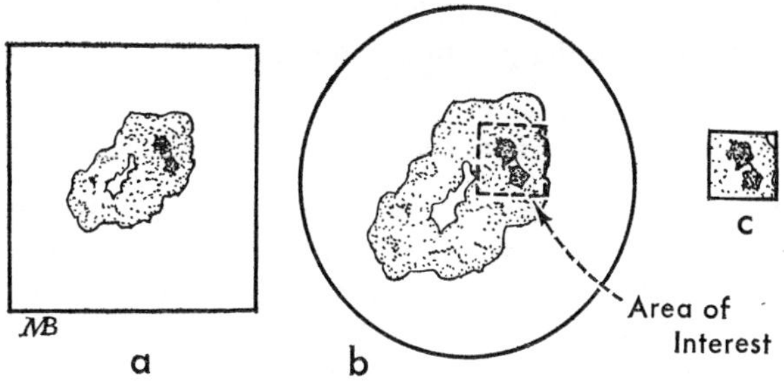

Fig. 11. Choosing an area of tissue for ultrathin sectioning. (*a*) Cut face of block, seen under stereomicroscope. (*b*) 1 μm section from block face, seen with 10 × objective. (*c*) Block face trimmed to area of interest.

double-edged stainless steel razor blade. (It is best to snap this blade in two lengthwise and grip the fractured edge when trimming, to avoid injuries.) The block should be trimmed into a four-sided pyramid with a flat top (Fig. 12). This flat top is the cutting

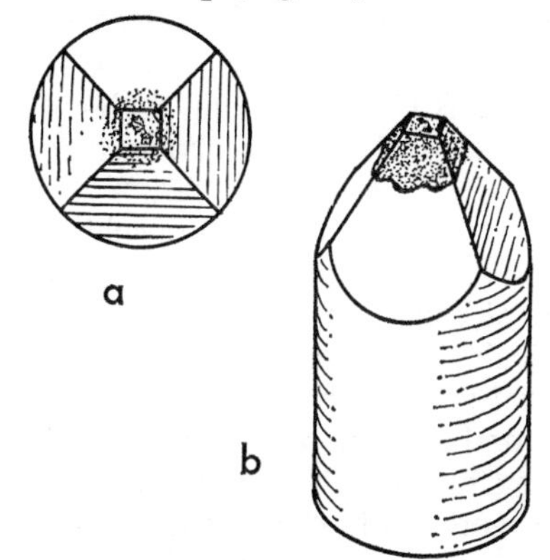

Fig. 12. The block trimmed for ultramicrotomy. (*a*) Top view. (*b*) Side view.

face. The pyramid furnishes a broad secure base for the cutting face and prevents the block from vibrating when passing the knife edge.

Some modern ultramicrotomes provide a special trimming holder which can be clamped onto the ultramicrotome itself. Trimming can thus be accomplished without the necessity of a separate stereomicroscope. The LKB ultramicrotome has a clever arrangement whereby the pyramid and cutting face may be fashioned with a glass knife. Also obtainable from some manufacturers is a special specimen trimmer to be used separately from the ultramicrotome. This is basically a stereomicroscope equipped with a specimen holder which is adjustable in all directions. It is fitted with a milling device for shaping the pyramid and cutting face.

ULTRAMICROTOMY

Once trimmed, the block is ready for ultramicrotomy. The detailed routine which must be followed in ultrathin sectioning varies from one type of microtome to another. Some microtomes are designed to be fully automatic once the block has been mounted and the appropriate controls set. The specimen is advanced toward the knife edge by the thermal expansion of a heated metal rod, and the fall of the specimen past the knife edge and its return stroke are controlled electrically. Supposedly all one has to do is to stand by and pick up the ribbons of serial sections as they come off the knife edge. In practice, however, it may take considerable dedication to master automatic ultramicrotomy.

Some manufacturers of automatic ultramicrotomes give regular courses of instruction in advanced ultramicrotomy. If one is going to spend a considerable portion of one's working life operating an automatic ultramicrotome, attending such a course is probably well worth it. The LKB Ultramicrotomy Laboratory in Stockholm gives 2-week courses which can be arranged in English, French or German.

Manual ultramicrotomy on a reliable machine such as the older models of the Cambridge Huxley will produce excellent single ultrathin sections, although obtaining ribbons of serial sections is usually more difficult than on an automatic machine. Learning to section on a manual machine has certain advantages. One of these is that the operator has a more intimate and immediate control over what is going on and the numerous variables involved in sectioning can be more easily checked through one by one. There is no doubt, however, that in sectioning difficult experimental tissues or tissues embedded in the water-soluble methacrylates the automatic ultramicrotome comes into its own. The new version of the Cambridge Huxley (Cambridge Mk II) is motorized with manual overide. It is reputed to provide the advantages of both manual and automatic sectioning.

Details of the operation of an ultramicrotome will be supplied by the manufacturer and demonstrated by his representative when the machine is installed. There are, however, some basic points to check which apply to all microtomes. All sources of possible vibration must be eliminated. Make sure that the block containing the tissue is securely placed in its holder and that the holder is firmly fastened to the microtome. Be sure that the knife is tilted at the proper angle and securely fixed in place. The knife tilt angle (the angle that the back of the knife makes with the vertical—also known as the clearance angle) should be somewhere between 1° and 7°. Check that the knife edge and the block face are parallel and so placed in relation to one another that the first cutting motion will not take off a large chunk of tissue. One thick section can dull the knife edge, and a very thick one will damage the cutting face of the specimen as well.

According to Porter (1964), if the block face has been trimmed so that it is longer than it is wide, the short edge should be positioned parallel to the knife edge. This allows the block to be moved along the knife edge more times before the good edge is finished. Sjöstrand (1967), however, recommends that the long edge of the rectangle parallel the knife edge.

The position of the block face, and all of the processes involved in sectioning can be observed at magnifications up to 100 times through a stereomicroscope mounted in front of or on the microtome. The illumination for this stereomicroscope should be adjustable and kept at an optimum.

Ten per cent acetone is placed in the boat on the front of the knife, and the meniscus adjusted to the proper level. As is the case when 1 μm sections are being cut, if the meniscus is too high (convex) acetone will leak over the back of the knife. Once this has happened it is very difficult to stop, as the surface tension at the knife edge has been broken, and the knife may have to be discarded. On the other hand, if the meniscus is too low (concave) sections will not float onto it as they come off the block.

It is wise to get the sectioning process started on the middle third of the knife and then when good sections start coming off, move the knife so that the best part of the edge is doing the cutting. Caution should be used when starting to cut after moving the knife laterally, since the knife may now be too close and may need to be backed off a little to avoid taking off a chunk of tissue.

The speed at which the specimen is moving when it passes the knife edge is important. This speed can be adjusted and will vary slightly with the type of tissue being cut. In general, a slow even rate of fall is desirable unless extremely thin (200 to 300 Å) sections are desired; then a more rapid fall is required.

All present-day ultramicrotomes have included in their design a method of assuring that the specimen does not pass too close to the knife edge on its return (upward) stroke. Sectioning the block face causes an electrical charge to build up on it. This charge plus capillary action of minute crevices in the block face is sufficient to draw liquid from the boat onto the block face, taking with it the section which has just been cut. On the next downstroke the water on the block face will wet the back of the knife, in all probability dragging the section down the back of the knife. Some ultramicrotomes prevent this by having the specimen arm swing to one side before starting the upstroke, so that the specimen does not pass the knife edge on its upward journey. The LKB ultramicrotome retracts the knife 20 to 30 μm before the upstroke, which gives adequate clearance.

Compression of the sections occurs during sectioning and is worse when the knife edge is dull or incorrectly oriented. Some of the compression is permanent and results in distortion of the tissue. Some of it is not permanent and can be reversed by holding a glass rod wet with chloroform above the sections. They can be seen to flatten out as the vapour reaches them.

Determining Section Thickness

Once the sections have been flattened, approximate section thickness can be determined by noting the colour of the sections as they reflect light. Silver-grey sections are approximately 500 to 600 Å thick and are what one should strive to obtain. Silver sections are approximately 600 to 900 Å thick; pale gold ones 900 to 1200 Å thick. The latter are certainly usable, although incapable of giving high resolution. Deep gold, purple, blue or green sections are too thick. All sections thinner than 500 Å are grey. They are also very rarely encountered by the beginner.

Section thickness can be varied by altering a setting on the microtome. The setting may have a scale associated with it which indicates section thickness, but this is at best a rough guide. The light reflected from the section is a truer indication of their thickness. For a discussion of this and other methods of assessing section thickness see Williams (1969).

Some Problems which may be Encountered while Sectioning

The number of sections which can be cut from one knife is determined by a number of variables which include the size of the block face, the thickness of the sections, the hardness of the tissue and the quality and length of the best part of the cutting edge. If score-marks or other defects appear on the sections, move to a fresh part of the edge. If a score-mark still appears in exactly the same position after the move, it may be something hard in the tissue itself which is causing the trouble. Possibly the tissue has been improperly fixed or embedded.

Horizontal striations (chatter marks) on the sections may be caused by the knife tilt being too great, or by the block not being trimmed in the correct pyramidal shape. More often, however, they are caused by the block not being firmly held in the chuck, the chuck not being firmly tightened onto the microtome, or the knife being loose in its holder. If everything has been properly tightened, vibration in or near the building may be causing the trouble. If these factors are ruled out, the microtome may be in need of adjustment or servicing.

A block of polymerized embedding material without tissue embedded in it is very helpful in diagnosing the source of trouble in ultramicrotomy. If a block with tissue in it is not cutting well, substitute the block with no tissue in it. If the latter cuts well, then the fault lies with the tissue. If the embedding material without tissue does not cut well either, the knife has a poor edge or is wrongly positioned, or the microtome is in need of servicing.

Picking up the Sections

After a ribbon of sections (or six or more separate sections) have been collected on the meniscus they are picked up on a grid. This is done by holding the grid by the edge with fine forceps and lowering it straight down onto the sections. Separate sections can be manoeuvred into position before the grid is lowered by means of an eyelash secured to an orangewood stick.

The copper grids normally used in routine electron microscopy are shiny on one side and have a matt finish on the other. It does not matter which side contacts the sections. However, one should be consistent about which side is used. The sections must always lie below the grid bars when the specimen is in place in the microscope column. Otherwise the magnification for a given setting on the magnification control knob will not be valid. Therefore, if sections are *always* placed on, say the matt side of the grid, one can tell at a glance how to orient the grid in the specimen holder. It is good practice for everyone using the microscope to adhere to a single convention regarding which side of the grid to use. This will avoid confusion.

After the grid has been touched to the sections, raise the grid from the meniscus. The sections will be floating on the drop of 10 per cent acetone which adheres to the grid. Turn the grid over so the drop is uppermost, and blot the underside on clean filter paper. The acetone will be absorbed into the filter paper, leaving the sections spread tightly across the grid bars.

Some people find it easier to put the grid beneath the surface of the meniscus and bring it up underneath the sections. Both methods work well, and choice is largely a matter of personal preference.

Occasionally difficulty is encountered in picking up the sections on the grid. The sections shoot away from the grid as if actively repelled from it. This may be a matter of electric charge. Earthing the microtome chuck which holds the specimen may help, or dipping the forceps in acetone before picking up the grid. If the grids are contaminated the sections may not pick up properly. If a vacuum pump is running in the same area where sectioning is being done, exhaust from the pump can contaminate exposed grids. A method of cleaning grids is given in the Appendix, p. 217.

There is little point in preparing more than four grids at one time from a block for routine purposes. The block should be saved so that more sections may be taken later at the same or a deeper level if need be.

Section Quality

There is no way of telling for certain what the quality of the sections will be without looking at them in the microscope. A section which appears perfect on the meniscus may contain tissue which is worthless when viewed in the electron microscope. On the other hand, a satisfactory photograph may be obtained of a section which appears somewhat less than perfect as it leaves the knife. Only experience will tell a person when a section is undoubtedly too poor to be worth picking up. If a tissue is extremely difficult to section it is worth looking at the occasional obviously poor section in the hope of gaining some information about the tissue. If the reason for the difficulty in sectioning is uncertain, a look at the sections in the electron microscope may indicate the cause of the trouble. This is particularly true of experimental tissue which has been subjected to unusual procedures.

Except in special circumstances it is best to stain the sections before looking at them. Assessing the quality of unstained tissue may be quite difficult, especially for a beginner. Secondly, the interaction of the electron beam with the section may render it partially resistant to staining. Thirdly, it has been stated (Shalla *et al.*, 1964) that staining will give higher contrast if done before the sections have thoroughly dried.

Sectioning for High Resolution Work

High resolution is arbitrarily defined in this book as a resolution better than 25 Å. A resolution of less than 15 Å has seldom been attained employing sections of embedded tissue, although Sjöstrand (1963) has reported a resolution of 10 Å and Haydon (1969) has reported 5 Å. The resolution obtainable is partially dependent upon the thickness of the section, and is usually considered to approximate a tenth of that thickness. (For a discussion of other factors involved see Wachtel *et al.*, 1966.) A minimum section thickness of 100 Å or less has been demonstrated by Sjöstrand (1967) and serial sections with an average section thickness of 200 to 300 Å are routinely obtained in his laboratory.

Most microscopists work with sections falling between 500 Å and 1000 Å in thickness. To obtain sections thinner than this, a very stringent adherence to certain rules of technique is necessary, as described by Sjöstrand (1967, 1969). A few of these rules are:

1. The knife edge must be of superior quality. It should be checked with good lighting at a high magnification.
2. The knife tilt angle should be small: 1°–3°.
3. The block face must be exceedingly small: $\frac{1}{10}$ mm square or less. The block should be so trimmed that the block face is supported by a pyramid to avoid vibration.
4. Cutting speed should be rapid.
5. Proper infiltration and polymerization of the embedding material is important to ensure maximum homogeniety in the block of tissue. Thinner sections are obtainable from hard blocks than from softer blocks.
6. The microtome must be functioning perfectly and be free from all vibration.

Since this is a beginner's handbook, only enough information on the fine art of ultrathin sectioning for high resolution electron microscopy is presented here to give an idea of the degree of perfection involved and the personal discipline necessary to obtain sections for high resolution work. When routine ultramicrotomy has been mastered to the point where silver or silver-grey sections are routinely produced, then it is time to read carefully the chapter by Sjöstrand (1967) on 'The Preparing of Thin Specimens for Electron Microscopy'. That chapter also gives invaluable advice on the art of serial sectioning.

Ultrathin Frozen Sections

A technique for obtaining ultrathin frozen sections of fixed or unfixed tissues is now under development in several laboratories, and is described in Chapter 8.

REFERENCES

Cameron, D. A. (1956). A note on breaking glass knives. *Journal of Biochemical Cytology* **2**, Supply., 57–59.

Fernandez-Moran, H. (1953). A diamond knife for ultrathin sectioning. *Experimental Cell Research* **5**, 255–256.

Haydon, G. B. (1969). An electron optical lens effect as a possible source of contrast in biological preparations. *Journal of Microscopy* **90**, 1–13.

Latta, H. & Hartmann, J. F. (1950). Use of a glass edge in thin sectioning for electron microscopy. *Proceedings of the Society for Experimental Biology and Medicine* **74**, 438–439.

Porter, K. R. (1964). In *Modern Developments in Electron Microscopy*, pp. 119–145 (Edited by B. M. Siegel), New York and London: Academic Press.

Pyper, A. S. (1970). A glass plier kit for making glass knives for electron microscopy. *Laboratory Practice* **19**, 491–492.

Shalla, T. A., Carrol, T. W. & De Zoeten, G. A. (1964). Penetration of stain in ultrathin sections of tobacco mosaic virus. *Stain Technology* **39**, 257–265.

Sjöstrand, F. S. (1963). A new ultrastructural element of the membranes in mitochondria and of some cytoplasmic membranes. *Journal of Ultrastructure Research* **9**, 340–361.

Sjöstrand, F. S. (1967). In *Electron Microscopy of Cells and Tissues*, Vol. 1, Chap. 9, pp. 222–310, New York and London: Academic Press.

Sjöstrand, F. S. (1969). In *Physical Techniques in Biological Research*, Vol. 3, Part C, Cells and Tissues, 2nd edn. pp. 139–200 (Edited by A. W. Pollister), New York and London: Academic Press.

Wachtel, A. W. *et al.* (1966). In *Physical Techniques in Biological Research*, Vol. III, Part A, Cells and Tissues. 2nd edn, Chap. 4 (Edited by A. W. Pollister), New York and London: Academic Press.

Williams, M. A. (1969). In *Advances in Optical and Electron Microscopy*, Vol. 3, pp. 228–238 (Edited by R. Barer and V. E. Cosslett), London and New York: Academic Press.

4

Staining of Biological Materials for Electron Microscopy

THE ORIGIN OF CONTRAST IN THE IMAGE

In electron microscopy the image of the specimen is formed by an electron beam focused by magnetic lenses. As the electron beam passes through the specimen, dense components in the specimen cause deflection (scatter) of some of the electrons in the beam. If deflection of the electrons by the tissue components is great enough, the electrons will fall outside the objective aperture and be completely eliminated from the beam. Thus a pattern will be formed upon a fluorescent screen or a photographic plate placed beneath the specimen in the path of the transmitted beam.

Dark areas on the fluorescent screen represent dense components of the specimen which have knocked electrons from the beam. Light (fluorescing) areas indicate where electrons have been able to pass through the specimen and energize the phosphor of the fluorescent screen. The amount of scattering occurring at any point in the specimen depends upon the thickness of the section and the atomic weights of its constituent atoms. If less than 5 per cent of the beam is scattered, the specimen is not visible to the eye (Zobel and Beer, 1965).

With low accelerating voltages the electrons in the beam have less energy and are more easily scattered. In the range of voltages commonly used in electron microscopy, contrast is inversely related to the accelerating voltage. If the objective aperture is small, more electrons will fall outside it, and again contrast is increased. The following factors, then, affect contrast in electron microscopy:

1. Section thickness
2. The presence of heavy atoms in the specimen
3. Accelerating voltage
4. Aperture size

CONTRAST ENHANCEMENT

Whereas in light microscopy tissue components are differentiated by staining with dyestuffs, the 'stains' in electron microscopy are compounds or ions containing elements of high atomic number which scatter electrons strongly. The larger the number of protons in an atomic nucleus, the larger the number of electrons which surround the nucleus. Both the atomic nucleus and the electron cloud around it deflect electrons in the beam passing through the specimen. However, scattering of beam electrons by the electron cloud (inelastic scattering) contributes most to the contrast of the image.

Most of the atoms in animal and plant tissue are of low atomic number:

Hydrogen	1
Carbon	12
Nitrogen	14
Oxygen	16
Phosphorus	31
Sulphur	32

The electron-scattering ability of these light elements is very small.

In an attempt to produce a sufficient range of densities in tissue components so that a satisfactory image may be formed, a number of heavy metals (i.e. atoms with large nuclei surrounded by a cloud of many electrons) have been used as 'electron stains'. A purely random distribution of heavy metal atoms throughout the tissue would not, of course, give meaningful contrast to the electron image. It is necessary to increase the mass of specific regions of the specimen. A number of heavy metal salts have proved useful as contrast-increasing electron stains, and about half a dozen are in common use.

As with the fixatives, the reactions of electron stains with tissue components are poorly understood. Most heavy metal cations will form complexes with negatively charged groups in the tissues, usually involving nitrogen, oxygen, sulphur and phosphorus. Ions of lead, uranium, indium, bismuth and iron will bind to oxygen, and thus may attach to the phosphate groups of nucleic acids (Zobel and Beer, 1965). Lead will also attach to osmium compounds bound to tissue components during fixation. Certain uranium salts (especially uranyl acetate) have fairly strong affinities for nucleic acids, but also act as general protein stains.

STAINING DURING FIXATION OR DEHYDRATION

Osmium tetroxide, since it includes the heavy metal atom osmium, stains as it fixes. It reacts with both lipid and protein. The contrast of lipoprotein cell membranes and fatty inclusions is intensified. Nucleoprotein contrast is increased slightly because of reaction of the osmium with its protein portion.

Potassium permanganate also stains as it fixes, enhancing the contrast of phospholipid membranes.

Staining can also be done by adding other heavy metal compounds to either the fixative itself or to the dehydrating fluids. Uranyl acetate and phosphotungstic acid have been used this way since the early days of electron microscopy. More recently, ruthenium red (Luft, 1966) has been found of value in enhancing the contrast of certain extracellular carbohydrate-containing compounds. It has been suggested (MacRae and Meets, 1970) that the increased contrast in nuclei brought about by soaking tissue in ammoniacal silver nitrate solution prior to dehydration is due to the staining of histones, but this has yet to be confirmed.

The chief drawback of staining during fixation or dehydration is that penetration of the stain into the tissue block is generally poor. Even when penetration into the extracellular spaces occurs, entry of the large stain molecules into individual cells is often prevented by the plasma membrane barriers.

STAINING OF ULTRATHIN SECTIONS

By far the most convenient and effective way of enhancing contrast has proved to be the staining of ultrathin sections of embedded material. Such sections, which are thin in comparison with the diameter of any cell, may be exposed on one or both sides to staining solutions. The stain now has no plasma membrane barriers to pass and can act directly on the exposed cell interiors. The only hindrance to passage of the stain is the embedding medium. Fortunately, certain reactions of electron stains with tissues embedded in epoxy resins and methacrylates can and do occur, so that the staining of tissue sections is a relatively simple matter. A grid bearing ultrathin sections is simply floated on or immersed in a drop of staining solution for an appropriate period of time. The embedding medium does not stain; the cut surface of the tissue does, as long as its reactive sites have not been blocked during processing.

Selectivity and Specificity

Results with most techniques in electron microscopy are notoriously variable from one laboratory to another, and staining results are no exception. Small alterations in the techniques of tissue preparation anywhere along the line can affect the affinity of the ultra-thin sections for a stain. Results will vary not only with the type and time of fixation, but also with the buffer employed as a medium for the fixative. The various embedding media differ from one another in their effects on the tissue. Epon-embedded tissue will generally stain more rapidly than Araldite-embedded tissue. Tissue embedded in water-soluble methacrylates will stain differently from tissue embedded in non-water-soluble methacrylates, and both will differ from tissue embedded in the epoxy resins. The temperature and the pH at which staining is carried out are also important. Nevertheless certain stains such as uranyl acetate and Reynolds' lead citrate when used under standard conditions will give fairly consistent results.

It should be remembered that the reaction which occurs upon contact of stain with section may not involve a cell constituent at all. The stain may react with a molecule of fixative which has attached to the tissue, as in the case of lead, which appears to attach to osmium compounds bound to certain cellular components during fixation (Marinozzi, 1963). Thus after osmium fixation lead acts as a general stain. The staining that is obtained with lead after aldehyde fixation alone appears to be largely confined to RNA-containing material (Marinozzi, 1963) and hydroxyl groups of carbohydrates (Daems and Persijn, 1963).

Several electron stains show a certain selectivity for various cell components, but none of them can be unequivocally termed a *specific* stain. If the binding sites of protein and carbohydrate are first blocked, indium trichloride will stain only DNA and RNA, but will not differentiate between the two (Watson and Aldridge, 1961). Uranyl acetate strongly enhances the contrast of both DNA and RNA, but also stains protein to a lesser extent. It is regarded as selective but not specific for nucleic acids. Under certain very strict conditions of staining, uranyl acetate has been reported to differentiate between RNA and DNA (Huxley and Zubay, 1961; Bernhard, 1969; Monneron and Bernhard, 1969).

Lead solutions at high pH (first introduced by Watson in 1958) impart excellent contrast to cell membranes. They also stain glycogen and nucleic acids. A number of workers, including Marinozzi (1963) and Daems and Persijn (1963) state that RNA and DNA may be differentiated by using uranyl acetate and lead under rigidly defined conditions. A complex cation containing two lead atoms $[Pb(OH)_2Pb]^{2+}$ appears to be the active ionic species in all the commonly used lead stains at high pH (Reynolds, 1963).

Phosphotungstic acid (PTA) below pH 2 has a high affinity for polysaccharides (Pease, 1966; Rambourg, Hernandez and

Leblond, 1969), but the specificity of the reaction is still hotly debated (cf. Glick and Stott, 1970; and Pease, 1970). At pH 3·0 to 3·5, PTA shows affinity for proteins and nucleoproteins.

In short, specificity (in a strict cytochemical sense) with the 'electron stains' now in use is at best controversial. More will be said about this in Chapter 8.

Stains in Common Use

Appendix VI gives recipes and routines for the most useful of the common stains. The beginner should familiarize himself thoroughly with the results obtainable by these methods before attempting to improve upon them. For a detailed discussion of these techniques and their variations see Pease (1964), Glauert (1955*a*), and the other references at the end of this chapter and Chapter 8.

Table 4 lists the most commonly used 'electron stains' and summarizes what little is known about their mode of action. It should be noted that the concentration of stain most commonly used is 1 per cent regardless of the atomic weight of the heavy metal. It has been found that for a specimen of given thickness the concentration of an electron stain required to increase contrast is independent of the nature of the stain used. When equal weights (i.e. equal percentages) are used, this in effect means that roughly equal numbers of electrons are present per millilitre in each stain. When equal molarities are used, however, contrast depends upon the molecular weight (Pearse, 1961).

Double Staining

One of the most commonly employed staining techniques is uranyl acetate followed by Reynold's lead citrate (Reynolds, 1963). Provided that uranyl acetate is used first, this procedure enhances specimen contrast more than does either stain used separately (Huxley and Zubay, 1961). Compare Plate 2(b), (c) and (d).

When the tissue has been fixed in osmium tetroxide, the uranyl acetate–lead citrate method would be more accurately described as a 'triple stain', since the osmium, uranium and lead all contribute to the increase in contrast. A procedure for this technique is given in the Appendix, p. 204.

Scanning Before Staining

There are certain circumstances in which one may wish to scan and photograph a grid both before and after staining. Such may

Table 4. Heavy metals commonly used to increase contrast in electron microscopy

Atom	Atomic weight	Staining solution used	Usual per cent concentration	Medium	Possible sites of attachment	Effects on image contrast: degree of enhancement
Osmium	190	Osmium tetroxide, stains during fixation	1%	Aqueous buffered solution; vapour if staining sections	C=C and polar groups of lipid, SH groups of protein	Protein + Lipid (phospholipid membranes, lipid inclusions, secretions) + +
Uranium	238	Uranyl acetate	1-2% (aq.); saturated sol'n in 50% ethanol	Water; 50% ethanol or methanol	PO_4 groups of nucleic acid. Also protein	Nucleic acid (chromatin, ribosomes) + + + Proteins + +
Lead	207	Lead at high pH + a chelating agent	About 1%	Water plus stabilizing agent and alkali	Reduced osmium, OH groups of carbohydrates, RNA	Lipoprotein membranes + + + Protein + Glycogen + + Ribonucleoprotein + + +
Tungsten	184	Phosphotungstic acid at pH 3 or above	1%	HC1 or chromic acid	Polysaccharides and glycoproteins (OH groups)	Mimics PAS reaction
		Phosphotungstic acid at pH 3 or below	1% in ethanol; 5 to 10% (aq.)	Absolute ethanol or water	Protein and nucleoprotein	Stains protein, especially collagen

be the case when a cytochemical reaction in the tissue has occurred which results in the deposition of lead at the reaction site. Obviously, one would wish to record the presence of the reaction product before staining with a lead stain, or any other stain which might obscure the reaction site. However, contrast in the specimen may be so poor without staining that it is difficult or impossible to identify what part of a cell, or what cell type, has reacted. Therefore after recording the reaction sites photographically it would be helpful to stain the sections and rephotograph the same areas.

In doing this it must be remembered that the electron beam changes the specimen in important ways. Ionization of specimen components by the electrons in the beam may change both its structure and its chemical composition. This, of course, may alter its affinity for the stain. Also, with long exposure to the beam a layer of contamination builds up on the specimen which can effectively insulate it from contact with the stain molecules. If viewing and photographing times are kept short, however, the adverse effects on subsequent staining may be kept to a minimum.

Importance of Stain Aggregate Size

The usefulness of a contrast-increasing stain in high resolution electron microscopy is limited in part by the size of the aggregate which attaches to the tissue components. A large aggregate of metal-containing stain will, of course, obscure any detail smaller than the dimensions of the aggregate. Silver is used only for special purposes because of the large size of the aggregate (40 to 50 Å) deposited in the tissue under the staining conditions ordinarily employed.

Sjöstrand (1963) reports a resolution of 10 Å on ultrathin sections after staining with uranyl acetate. In another paper Sjöstrand (1969) states that lead staining may result in deposition of stain clusters measuring up to 50 Å in diameter. He does not, however, specify under what conditions such large aggregates are deposited. Haydon (1969), using the popular Reynolds' lead citrate stain, reports lead clusters measuring just over 10 Å in diameter.

Many factors influence the size of the stain aggregates formed. These include, among others, the identity and concentration of the heavy metal atoms used to effect staining, the presence of other ions in the staining solution, temperature and pH of the stain, and length of staining time. The type of fixative used on the tissue and steric relationships of the tissue components to which the stain will attach may also influence aggregate size. The interaction of these various factors is not well understood. Stain aggregate size for particular stains under specified conditions does not appear to have been investigated in a systematic manner. Therefore, the art of electron staining for high resolution still requires a great deal of empirical juggling.

SURFACE STAINING

A limiting factor in the resolution obtainable with ultrathin sections is specimen thickness. Scattering of electrons by a thick section can produce sufficient chromatic aberration to limit resolution, as explained in Chapter 1. Superposition of structures within the section also limits resolution.

If bodies measuring 50 Å in diameter lie closely packed together in several planes within a section, they will appear superimposed upon one another if the section is more than 50 Å thick (Sjöstrand, 1969). Since the depth of focus of the electron microscope is much greater than the section thickness, all of the bodies in the section will appear to be in focus at the same time. This results in an uninterpretable jumble of fine structure. Cosslett calculated some years ago that the resolution obtainable on a section was roughly one tenth of the section thickness.

In practice, it is seldom possible to cut sections thinner than 100 to 200 Å, and routine sections obtained in most laboratories are in the silver-grey to pale-gold range which represents a thickness of 500 to 900 Å. Therefore for higher resolution work 'surface staining' of the section is usually recommended.

In theory, if the grid bearing the ultrathin sections is floated section-side down on a drop of stain rather than being immersed in the fluid, only the surface of one side of the section will be stained. It has been assumed that the stain does not penetrate through the section, since staining only one side results in less contrast than does immersing the section. If osmium tetroxide has been used as a fixative, material throughout the section will, of course, have a certain amount of contrast. Nevertheless, if the contrast added to the surface by the stain is great enough, the unstained material beneath the surface will be so pale in comparison that its contribution to the image will be negligible (Sjöstrand, 1969). By surface staining with Reynold's lead citrate a resolution in the region of 10 Å on a 500 Å Epon section has been reported (Haydon, 1969). It appears, however, that the success of surface staining is dependent on a number of variables. Shalla, Carroll and de Zoeten (1964) have shown that if sections of tissue embedded in Araldite or Maraglas (an epoxy resin) are not allowed to dry after cutting, penetration of strontium permanganate stain occurs where tissue components intersect the surface of the section. They reported that the stain did not, however, penetrate through pure epoxy resin to reach structures completely enclosed therein. More recently, Peters, Hinds and Vaughn (1971) have presented convincing evidence that both uranyl acetate and lead citrate stains completely penetrate sections of Araldite 502 up to 2000 Å in thickness, and stain small particles such as ribosomes which are entirely enclosed within the section.

Obviously, a series of carefully controlled experiments to determine the depth of penetration, under a variety of conditions, of the common electron stains into sections of tissues embedded in different media would be a valuable contribution to this field. Meanwhile one can infer from the work of Shalla, Carroll and de Zoeten (1964) that if a purely surface stain for high resolution work is desired it is more likely to be attained if the sections are thoroughly dried before staining. If maximum contrast is desired at the expense of high resolution, immersion of wet sections would appear to be preferable, possibly coupled with methods which reputedly increase stain penetration (Pease, 1964).

Negative Staining

This technique is used with particulate biological material and does not apply to ultrathin sections.

The attachment of heavy atoms to specimen components is known as 'positive staining', since the specimen itself is stained. However, if a biological specimen is immersed in an electron-dense material which does not stain the specimen but which creates a dark ground around it, it is said to be negatively stained. The dense material may also penetrate into crevices or lie in the contours of a specimen, thereby causing surface detail to stand out in contrast. A routine procedure for this technique was worked out by Brenner and Horne (1959), and has yielded much useful information regarding the structure of bacteria, viruses, isolated cell organelles, fibres, filaments and macromolecules.

A number of heavy metal salts have been used as negative stains, but by far the most useful has been sodium phosphotungstate. A suspension of the material to be studied is usually made in one per cent aqueous sodium phosphotungstate. The suspension is then either vaporized in a nebulizer and sprayed on the surface of a collodion- or Formvar-coated grid, or applied with a micropipette. Drying is almost instantaneous, and a thin film of stain incorporating the specimen is produced.

A number of artefacts may be introduced by the techniques involved in negative staining. For instance, isolated cell components are put (usually without prior fixation) into the negative stain after having undergone a number of sequential procedures to separate them from their normal environment. Details of the interaction of the negative stain with these components are uncertain. The drying and subsequent exposure to the electron beam can also be expected to affect the ultrastructure.

The resolution obtainable with negative staining is greater than that generally obtainable with ultrathin sections. When phosphotungstate is used, the limit is approximately 12 Å, which is the size of the granules of the dried stain (Parsons, 1965). Slightly better resolution is obtainable with finer grain negative stains such as uranyl formate (Leberman, 1965).

For detailed discussions of negative staining methods see Horne (1965*a* and *b*), and for factors influencing the appearance of biological specimens in negatively stained preparations see Glauert (1965*b*).

REFERENCES

Staining during tissue processing

Luft, J. H. (1966). Fine structure of capillary and endocapillary layer as revealed by ruthenium red. *Federation Proceedings. Federation of American Societies for Experimental Biology* **25**, 1773–1783.

MacRae, E. K. & Meetz, G. D. (1970). Electron microscopy of the ammoniacal silver reaction for histones in the erythropoietic cells of the chick. *Journal of Cell Biology* **45**, 235–245.

Staining of ultrathin sections

Beer, M. (1965). Selective staining for electron microscopy. *Laboratory Investigation* **14**, 1020–1025.

Bernhard, W. (1969). A new staining procedure for electron microscopical cytology. *Journal of Ultrastructure Research* **27**, 250–265.

Daems, W. Th. & Persijn, J.-P. (1963). Selective staining with heavy metals of osmium-fixed and formol-fixed mouse liver. *Journal of The Royal Microscopical Society* **81**, 199–201.

Glauert, A. M. (1965*a*). In *Techniques for Electron Microscopy*, Chap. 9 (Edited by D. H. Kay), Oxford: Blackwell Scientific Publications.

Glick, D. & Scott, J. E. (1970). Phosphotungstic acid not a stain for polysaccharide. *Journal of Histochemistry and Cytochemistry* **18**, 455.

Haydon, G. B. (1969). Electron phase and amplitude images of stained biological thin sections. *Journal of Microscopy* **89**, 73–82.

Huxley, H. E. & Zubay, G. (1961). Preferential staining of nucleic acid-containing structures for electron microscopy. *Journal of Biophysical and Biochemical Cytology* **11**, 273–296.

Marinozzi, V. (1963). The role of fixation in electron staining. *Journal of the Royal Microscopical Society* **81**, 141–154.

Monneron, A. & Bernhard, W. (1969). Fine structural organization of the interphase nucleus in some mammalian cells. *Journal of Ultrastructure Research* **27**, 266–288.

Pearse, A. G. E. (1961). In *Histochemistry, Theoretical and Applied*, 2nd edn, pp. 767–768, London: J. & A. Churchill.
Pease, D. C. (1964). *Histological Techniques for Electron Microscopy*, 2nd edn, Chap. 7, New York and London: Academic Press.
Pease, D. C. (1966). Polysaccharides associated with the exterior surface of epithelial cells: kidney, intestine, brain. *Journal of Ultrastructure Research* **15**, 555–588.
Pease, D. C. (1970). Phosphotungstic acid as a specific electron stain for complex carbohydrates. *Journal of Histochemistry and Cytochemistry* **18**, 455–458.
Peters, A., Hinds, P. L. & Vaughn, J. E. (1971). Extent of stain penetration in sections prepared for electron microscopy. *Journal of Ultrastructure Research* **36**, 37–45.
Rambourg, A., Hernandez, W. & Leblond, C. P. (1969). Detection of complex carbohydrates in the Golgi apparatus of rat cells. *Journal of Cell Biology* **40**, 395–414.
Reynolds, E. S. (1963). The use of lead citrate at high pH as an electron-opaque stain in electron microscopy. *Journal of Cell Biology* **17**, 208–212.
Shalla, T. A., Carroll, T. W. & de Zoeten, G. A. (1964). Penetration of stain in ultrathin sections of tobacco mosaic virus. *Stain Technology* **39**, 257–265.
Sjöstrand, F. S. (1963). A new ultrastructural element of the membranes in mitochondria and some cytoplasmic membranes. *Journal of Ultrastructure Research* **9**, 340–361.
Sjöstrand, F. S. (1967). In *Electron Microscopy of Cells and Tissues*, Vol. 1, p. 391, New York and London: Academic Press.
Sjöstrand, F. S. (1969). In *Physical Techniques in Biological Research*, Vol. 3, Part C, pp. 169–200 (Edited by A. W. Pollister), New York and London: Academic Press.
Wachtel, A. W., Gettner, M. E. & Ornstein, L. (1966). In *Physical Techniques in Biological Research*, Vol. 3, Part A, pp. 239–246 (Edited by A. W. Pollister), New York and London: Academic Press.
Watson, M. L. (1958). Staining of tissue sections for electron microscopy. II. Application of solutions containing lead and barium. *Journal of Biophysical and Biochemical Cytology* **4**, 727–730.
Watson, M. L. & Aldridge, W. G. (1961). Methods for the use of indium as an electron stain for nucleic acids. *Journal of Biophysical and Biochemical Cytology* **11**, 257–272.
Zobel, C. R. & Beer, M. (1961). Electron Stains. I. Studies on the interaction of DNA with uranyl salts. *Journal of Biophysical and Biochemical Cytology* **10**, 335–346.
Zobel, R. E. & Beer, M. (1965). The use of heavy metal salts as electron stains. *International Review of Cytology* **18**, 363–400.

Negative staining

Anderson, T. F. (1962). In *The Interpretation of Ultrastructure*, pp. 251–262 (Edited by R. J. C. Harris), New York: Academic Press.
Brenner, S. & Horne, R. W. (1959). A negative staining method for high resolution electron microscopy of viruses. *Biochimica et biophysica Acta* **34**, 103–110.
Glauert, A. M. (1965*b*). Factors influencing the appearance of biologic specimens in negatively stained preparations. *Laboratory Investigation* **14**, 1069–1079.

Horne, R. W. (1965*a*). The application of negative staining methods to quantitative electron microscopy. *Laboratory Investigation* **14**, 1054–1068.

Horne, R. W. (1965*b*). In *Techniques for Electron Microscopy*, Chap. 11 (Edited by D. H. Kay), Oxford: Blackwell Scientific Publications.

Leberman, R. (1965). Use of uranyl formate as a negative stain. *Journal of Molecular Biology* **13**, 607–610.

Parsons, D. F. (1965). Effects of the preparation procedures on the appearance of the object in electron micrographs. *Laboratory Investigation* **14**, 1169–1177.

Valentine, R. C. (1959). The shape of protein molecules suggested by electron microscopy. *Nature, London* **184**, 1838–1841.

Valentine, R. E. & Horne, R. W. (1962). In *Interpretation of Ultrastructure*, pp. 263–278 (Edited by R. J. C. Harris), New York: Academic Press.

5

Routine Operation of the Electron Microscope

Certain general aspects of operation are common to most modern electron microscopes and will be discussed in this chapter. Details of design and operation vary greatly from one model to another; many operations are fully automatic on the newer instruments. Operation of a particular instrument can best be learned under the supervision of an experienced operator. After initial briefings, the microscope may be operated with the aid of the instruction manual provided by the manufacturer. Some instruction manuals provide outlines of routine procedures on sheets which can be removed from the manual and taped in position for easy observation.

Handling of small and delicate items associated with electron microscopy such as specimen supporting grids, specimen holders, apertures and aperture holders will be difficult at first. When possible (i.e. when a safelight is not required) work with adequate lighting—if necessary with the aid of a dissecting microscope. Fine-pointed tweezers are essential for manoeuvring grids, and should be clean and in perfect condition.

Since the electron microscope is operated in semidarkness, a great many manipulations are done by feel. An operator soon develops his kinaesthetic sense so that his hands automatically fall on the proper controls when he needs to use them. This process can be speeded up by a little practice with the lights on. Form a mental picture of the control panels and of the various devices on the column which you will need to find in the dark. Practice finding them without looking, and become familiar with those controls which are close enough to them to be confused with them. For instance if electron gun tilt controls are close to the beam centring controls, confusion of these in the dark can result in misalignment of the illumination and probably disappearance of the electron beam. It is wise to make a note of the usual settings of the various controls in case they are altered by mistake.

It is helpful when one is first becoming accustomed to the controls to put a white spot on those which are used constantly, such as the fine focus control for the second condenser lens (hereafter referred to as condenser 2 or C2) and the beam centring controls. This spot will be visible in the semidarkness and will help to avoid confusion. The round white adhesive spots used for orientation of lantern slides are convenient to use for this purpose.

Critical viewing of a specimen on the fluorescent screen and accurate focusing of the image require that the eyes have become dark-adapted and remain so during microscope operations. For optimum results the safelight should be off and the panel lights dimmed to their lowest value or extinguished. Under these conditions the fluorescent screen will appear brighter and have more contrast. This is important since both focusing and photographic exposure should be done at the same relatively low beam intensity on the fluorescent screen in order to attain optimum exposure without specimen drift.

TYPICAL OPERATIONS INVOLVED IN A DAY'S ELECTRON MICROSCOPY

In the following description it is assumed that the instrument has been turned on and aligned according to the manufacturer's instructions, and that the astigmatism of the condenser and objective lenses has been compensated using a carbon film with holes, as described on pp. 92–94. The accelerating voltage (sometimes referred to as high voltage, high tension, or HT) is on, but the filament current is off and the machine is in the idling condition specified in the instruction manual.

Putting the Specimen into the Microscope

When not in the microscope, specimen holders should be kept in a clean covered jar containing a desiccant. This will prevent water vapour from entering the microscope column on the specimen holder.

At no time should the holder be touched by the fingers, since grease and moisture from the fingers will contribute to the contamination of the microscope column (see p. 90). Handle the grid with clean tweezers and the specimen holder with white nylon gloves or lint-free vellum, unless a special handling device is provided.

The grid bearing the specimen is inserted in the specimen holder so that the sections will face downward in the microscope column.

(It is thus important always to place your specimens on the same surface of the grid when mounting them, as described in Chapter 3, so you will be certain to know which side they are on.) The sections should face downward to ensure the proper position of the specimen relative to the objective lens field. The magnification produced by the objective lens varies with the position of the specimen on the microscope axis.

The specimen holder may now be inserted into the microscope specimen chamber. In order to do this, the specimen chamber (sometimes referred to as the specimen airlock) must first be closed off from the rest of the microscope column by closing the appropriate valve. Then air is leaked into the specimen chamber. When the chamber has reached atmospheric pressure the door of the chamber may be opened. Once the specimen holder is in position in the specimen chamber, the door is closed and the chamber is evacuated. This usually takes about thirty seconds. The evacuation can be followed either by watching a Geissler tube (sometimes called an ionization gauge) or a pressure indicator meter. A Geissler tube is a fluorescent tube on the control panel of some machines. It glows pink when air is in the specimen chamber, glows blue as the air pressure diminishes, and finally goes dark when the required vacuum has been attained. Then the specimen chamber may be opened to the rest of the column.

Obtaining Proper Illumination

After the specimen has been inserted and the chamber evacuated and opened to the column, the filament current may be turned on. If the microscope has been properly aligned and compensated, an electron beam will appear on the fluorescent screen as the filament current is increased.

At first several 'spots' of illumination will be seen on the screen. These correspond to the several foci of electron emission from the surface of the filament which occur at low current flux. As the current is increased, emission occurs only in the region of the filament tip. The bias voltage on the cathode shield (which is negative with respect to the filament) condenses the cloud of emitted electrons to a dense concentration below the tip of the filament. This concentration acts as the effective electron source. A single spot of illumination is now seen on the fluorescent screen. The filament current should be increased only until maximum brightness of the focused spot is attained. Increasing the current beyond this point will cause no increase in brightness and will shorten the life of the filament.

When maximum illumination is attained, the beam is centred on the screen by manipulation of the beam centring controls. These controls adjust the illumination axis with respect to the objective lens. Then the beam may be spread (defocused) to illuminate the desired area by using the condenser 2 fine focusing control. The illumination should spread evenly and concentrically over the field if the condenser aperture is correctly centred and the condenser astigmatism compensated for.

What to do if a beam is not obtained

If the beam does not appear on the fluorescent screen when the filament current is turned up:

1. Make sure that the high voltage it switched on.
2. Check the filament current monitor meter. If the needle does not go up as the current control is turned up, usually the filament has open-circuited ('blown') and no current is going through the circuit. If this is the case, the filament must be replaced (see Chapter 11, p. 184).
3. If the current monitor meter shows that current is flowing through the filament, check to make sure that the specimen chamber is open to the column and that the specimen holder is correctly in place.
4. Turn the specimen traverse controls to make sure that a grid bar is not obscuring the illumination.
5. Move the objective aperture control to the 'aperture out' position so that the aperture is no longer in the beam path. If the beam appears, the objective aperture has been obscuring the beam and must be recentred.
6. Check that all the control settings are in the idling positions described in the instruction manual.
7. Open the specimen chamber as described previously and remove specimen holder. Make sure you have not put your grid in on top of someone else's in the holder.
8. Remove the specimen and reinsert the empty specimen holder into the specimen chamber. Pump out the chamber and again try to get a beam.
9. If all these things appear to be in order and there is still no beam, the routine daily alignment procedure described in the instruction manual should be carried out.

Locating an Area of the Specimen for Study

Some microscopes have a special low magnification setting for locating sections of the specimen on the supporting grid. This saves the trouble of laboriously scanning the grid hole by hole at a magnification of 1000 times or more. If there is no 'scanning setting' it is advisable to scan the grid starting at one side and working systematically toward the other. This is easier with slot grids, i.e. grids with long rectangular apertures, especially if the lowest magnification setting allows two slots to be scanned at the same time. If you form the habit of always working in the same direction, you will not miss any important material and it will be easier to remember where you were if you are interrupted.

If time permits, it is wise to scan a grid and photograph anything you think important on it at one go. It is notoriously difficult to re-locate a particular area on a specimen once you have removed the grid from the microscope. Also there are the hazards of losing or damaging the grid during handling.

When a section of specimen has been located on the grid, increase the magnification until you can recognize the gross components of the specimen (e.g. individual cells or cell nuclei). The beam may have to be recentred using the beam centring controls when the magnification setting is changed. The best magnification for location of areas of interest will vary with the nature of the specimen and with the contrast which has been attained by staining. In specimens of very low contrast, it may be impossible at the lowest magnification to be sure whether the sections contain tissue or not. Increased magnification and use of the binocular viewer provided with the microscope will aid you in locating pertinent material. Be sure that the interocular distance of the binocular viewer is properly adjusted for your eyes, and that the viewer is focused on the grain of the fluorescent screen.

WHAT TO PHOTOGRAPH

The place to study an electron image is not the fluorescent screen. The resolution of most phosphors is limited to about 50 to 100 μm. Moreover, unless the microscope is fitted with an anti-contamination device the specimen can become contaminated so rapidly in the electron beam that detail is soon lost. In order to preserve important detail, it is essential to photograph it.

What, then, is important detail? This depends on the type of tissue and what the operator wishes to learn from his study. If the tissue concerned has received little study at the electron microcopic level, and you wish to describe its ultrastructure fully, you have not much choice but to photograph extensively grids taken from areas which have been carefully selected by studying orientation slides. Only by making a fairly thorough photographic survey of suitable areas will you amass sufficient detailed information to be of value. On the other hand, if the ultrastructure of the tissue is already fairly well known and you wish only to study a particular structure or one cell type, then the problem of what to photograph is much simpler. You have only to scan grids until you find that structure or that cell type.

When you have found what you wish to photograph, do so as quickly as is compatible with accurate focusing. The longer you take to photograph an area the more contamination accumulates on the specimen and the less detail will be recorded. Also the electron beam may cause damage to the specimen (see p. 91). It is usually good practice to take one or two low magnification photographs of an area to give the necessary orientation for later study. Unless an anti-contamination device is being used, it is better to do high magnification photography first in order to record the necessary fine detail before contamination builds up on the specimen. Then reduce the magnification to record a larger field.

Before photographing, check that the cross section of the electron beam is circular as it is spread across the fluorescent screen with the condenser 2 fine focus control. Otherwise, the field will not be illuminated evenly at high magnification. If the beam cross section is not round, compensate for condenser astigmatism as described on p. 92.

When setting the magnification one should always first increase the current in the projector lenses to a setting higher than that at which the photograph is to be taken. Then lower the projector current to attain the desired magnification setting. The reason for this is a phenomenon known as magnetic hysteresis. The field of a lens in which the current has been increased to a given magnification setting is not exactly the same strength as the field of the same lens at the same setting after the current has been decreased from a higher setting. In other words, the size of an image photographed after increasing the magnification from 16,000 to 20,000 will not be the same as the size of the image photographed after decreasing the magnification for 40,000 to 20,000. The difference in size can be as much as ± 10 per cent. It has been found that the magnification at any given setting is more reproducible when the lens field has just been decreased.

A few microscopes provide a special control so that hysteresis effects may be eliminated by pushing a button. This makes the preliminary increase and decrease of the projector lens current unnecessary.

Focusing

Accurate focusing requires practice. Often, particularly at high magnifications, one is faced with the problem of deciding which degree of fuzziness represents true focus. All that can be usefully given in this discussion is a few general suggestions. After that, it is a matter of doing a great deal of looking and photographing and critically assessing the results.

1. When focusing, use the binocular microscope for viewing the fluorescent screen. This microscope is installed so that it can be swung in and out of position when needed. Be sure it is properly adjusted for your eyes and is accurately focused on the grain of the fluorescent screen.
2. Always start with all the focusing knobs in approximately the middle of their runs.
3. Get as close as possible to focus by bringing the *coarsest* focusing knob back and forth through focus, i.e. from a position where it is obviously underfocused (in the anticlockwise direction; lens current reduced) to a position where it is obviously overfocused (in the clockwise direction; lens current increased).
4. Repeat this with each focusing knob, working downward from coarse to fine. You cannot get in focus by manipulating a fine focus control if the coarser controls are not near focus.
5. At lower magnifications (1000 to 4000) you will find that there is a considerable distance over which you can move the fine focusing knob without the image appearing to change. That is, the image will appear to be in sharp focus over a considerable portion of the knob's run. In actual fact, the image is underfocused over all of this distance except at the point just below which the image is seen to go obviously over focus. Therefore, turn the focusing knob clockwise until the image is first seen to be over focus. Then turn back just enough to make the image appear focused. A photograph taken at this point will be in or very close to focus. Because of phase contrast effects, a slightly underfocused image will appear sharper than does one in true focus, so considerable finesse in viewing must be developed, especially at low magnifications. At the very lowest magnifications, the fine focusing knob may cause no change in appearance of the image over its entire run. If so, use the next coarser knob in the same manner as described above.

At magnifications of approximately 6000 to 20,000, rotate the fine focus knob so that the image goes through focus and estimate the point which lies halfway between the just-visibly-underfocused and the just-visibly-overfocused positions. After some practice you will probably note that the position corresponding to true focus corresponds to the position where the specimen has the least contrast. (This is also true at both higher and lower magnifications, but at lower magnifications it only becomes evident if the objective aperture is removed. On some microscopes it is quite easy to remove the aperture when focusing and get it back into position quickly and easily; in other machines the difficulty involved makes this procedure rather impractical.)

At high magnifications (40,000 to 250,000) when one is working near the maximum resolving power of the instrument, diffraction and phase contrast effects again create problems in determining focus (Haydon, 1968; Sjöstrand, 1969). When the specimen is slightly underfocused, Fresnel (diffraction) fringes appear around structures in the specimen. These fringes increase the contrast of the image and may give the spurious appearance of fine detail (defocus granularity). Also structures appear larger than they are due to the presence of the fringe. If there is a small hole in the field of view, one can focus on the edge of it. When the Fresnel fringe at the edge of the hole disappears, focus has been attained. If no such convenient hole is present, it is necessary to take a through-focus series of photographs in order to assure getting one in true focus.

Focusing with Minimum Contamination of the Specimen

If you are using an instrument which is not fitted with an anti-contamination device and you have found a structure which you wish to photograph, it is important to do so quickly in order to avoid build-up of contamination on the specimen. This is particularly true at high magnifications. It is good practice to move away slightly from the area which you wish to photograph and focus upon another structure. Then come back to the area of interest. It will then only be necessary to correct the finest focusing knob before taking the photograph.

The finest focusing knob *must* be checked before each photograph, as neighbouring areas of a specimen may lie at slightly different planes due to sagging of the specimen from the grid bars.

Taking the Photograph

If the instrument is not equipped with an exposure meter it will be necessary to judge the proper amount of illumination for the micrograph by eye. This is rather difficult and requires practice. The degree of success varies considerably from one operator to another. Ideally, photographs should be taken with the same amount of illumination which was used for focusing. Decreasing the beam brightness after focusing may cause thermal drift (lateral movement) in the specimen, thereby reducing resolution. Focusing should be done with the minimum illumination compatible with accuracy. If your eyes are dark adapted and dial lights on the microscope are off, focusing can be accurately accomplished at the reduced illumination level which produces an optimal photographic exposure in a reasonable time.

Using Ilford EM6 plates it has been found that a 4-second exposure at reduced illumination produces a plate which has recorded the desired information and will develop to optimum contrast in a developer such as PQ Universal within 4 to 5 min. A little experience will enable you to judge the proper amount of illumination to produce this result. Exposure times in excess of 4 seconds are not recommended, as instabilities in the microscope or the specimen may cause blurring.

The manner of taking a photograph varies from one instrument to another, but usually involves introducing a glass plate or cut film into position under the fluorescent screen, or winding a section of roll film into this position. In some microscopes exposure is made by simply raising the fluorescent screen for the desired number of seconds and then lowering it again. This has the disadvantage that unless great care is taken vibration may be caused by the raising and lowering of the screen. A better method is to have an automatic shutter so that exposure is accurately timed and does not commence until vibration from raising the screen has ceased.

To avoid vibration, the operator should not lean on the microscope or rest his elbow on it during photography. Observers should be discouraged from this practice as well.

After the photograph is taken the exposed plate is returned into the camera cassette, or the film wound on. The details pertinent to the photograph such as plate number, identification of the structure photographed and the magnification should be recorded in a book kept for this purpose.

Removing Photographic Material from the Microscope and Reloading the Camera

When all the photographic material in the camera has been exposed, defocus C2, turn down the beam current, switch off the filament, and return the magnification to its lowest setting. Isolate the camera from the rest of the column by closing the appropriate valve. Air may then be admitted to the camera through an air-leak valve. The camera can then be opened and the plates or film removed. Some camera cassettes are light-tight and the camera can be opened with the lights on. Otherwise a safelight must be employed until the plates or film are put into a light-tight container.

Reload the camera with fresh plates or film (which have been kept dry in a vacuum desiccator) and pump out the air from the camera. While the camera is being pumped out, the exposed plates or film can be taken into the darkroom and developed. Take care that both your hands and the bench upon which you are working are clean and dry. Any contamination which gets on the plate or film trays may be transferred into the microscope column.

If plates are used they can be marked at one edge (on the emulsion side) with pertinent data such as the plate number and date. An ordinary graphite pencil will serve for this purpose. Make sure that the numbering on the plates corresponds with that in the log book. Roll film will probably already be numbered. After marking the plates, remove them from their trays and place them in a developing rack. Then process as directed in the chapter on darkroom procedures.

When the photographic material has been developed, fixed and put to rinse, check the Geissler tube or pressure indicator meter on the electron microscope. If it indicates that a sufficient vacuum has been attained in the camera (roughly between 10^{-1} and 10^{-2} mmHg) the camera can be opened to the rest of the column. Then, wait until the vacuum gauge indicates a vacuum of at least 10^{-4}mmHg, at which time the microscope is ready for operation and the filament may be switched on again.

While the microscope is completing this final pumping down, fresh plates can be loaded into the empty trays in the darkroom, or fresh roll film inserted into the spare camera magazine. The trays first should be carefully brushed out and treated with a jet of air to ensure that dust and glass particles are removed. *In no event should the trays or plates be handled with wet hands.*

When loading unexposed plates into the individual trays, the plates must be so placed that the emulsion side of the plate will face upward in the column. The emulsion and non-emulsion sides may be identical in colour. The non-emulsion side is the shinier of the two, while the emulsion side has a more nearly matt finish.

After the plates have been loaded or fresh roll film inserted into the spare magazine, they should be placed in a vacuum desiccator to remove moisture from both the photographic material and its container. Desiccation should proceed for at least 2 hours before the plates are placed in the microscope. Gelatin-halide emulsions have a high water content which must be removed insofar as possible to avoid water vapour molecules escaping into the column from the camera.

Importance of Rechecking the Astigmatism Correction

If accurate work at high magnification is being done, it is very important to make sure that the astigmatism of the objective lens remains corrected. In a clean microscope if the objective aperture is properly centred the chief source of increasing objective astigmatism is the buildup of contamination on the edge of the aperture. This aperture of necessity must be small (25 to 50 μm in diameter) to keep spherical aberration at a minimum. Some compensation for the contamination can be attained by using the objective astigmatism corrector. Nevertheless the aperture may have to be replaced after a time due to excessive buildup of contamination which cannot be compensated for. If manipulation of the objective astigmatism corrector (see p. 92) will not correct the astigmatism, remove the contaminated objective aperture and replace it with a clean one, following the directions in the instruction manual.

COMMON PROBLEMS ARISING DURING OPERATION OF THE MICROSCOPE

1. *Specimen Movement*

When the beam initially heats the specimen, movement may occur due to thermal effects. This is especially likely if the ultrathin sections are torn or not firmly in contact with the specimen supporting grid. It sometimes stabilizes the specimen if the beam is passed along the line of contact between specimen and grid bar, thus welding the section to the grid bar. If thermal drift is ruled out, the specimen stage may be defective and in need of attention.

2. *Insufficient Illumination*

If illumination falls off sharply as magnification increases, the illumination system may need adjustment. In some machines this may be simply a matter of manipulating the anode centring controls until illumination on the screen reaches a maximum. If the illumination system is correctly aligned and there is still insufficient illumination at higher magnifications, it may be necessary to reduce the bias voltage. If this is done, however, the filament saturation point should be rechecked so as not to run the filament at a higher beam current than necessary. If one is working at a low accelerating voltage and the lack of illumination is due to a thick specimen, increase the accelerating voltage. After increasing the voltage recentre the illumination, recheck the filament saturation point, and recheck the objective astigmatism correction.

3. *Elliptical Beam Cross Section*
This is due to condenser astigmatism, which should be corrected (see p. 92).

4. *Condenser Sweep*
If the beam sweeps across the field instead of expanding evenly about the screen centre, the condenser aperture needs re-centring. This is done by alternately manipulating the condenser aperture centring controls and the beam centring controls until the beam expands properly.

5. *Image Will Not Focus*
If sharp focus cannot be obtained with the fine focusing knob, check to see that the coarser focusing knobs have all been focused properly. Then ascertain that the astigmatism of the objective lens is still corrected. Then close off the valve to the specimen chamber, open the chamber and check to see that the grid is flat, sections downward, and properly placed in the specimen holder, and the holder properly seated. If none of these manoeuvres results in obtaining focus, a fault may be present in the electrical circuits.

6. *The Beam Disappears*
Check first to see if the high voltage has switched off due to incorrect operation of the microscope or a fault condition. If so, perform an emergency shutdown and ascertain what is wrong, e.g. water pressure to lens cooling system too low, air leak in the column, etc. If the high voltage is still on, check the filament current meter. If the needle has fallen to zero it is likely that the filament has blown. If the meter gives the normal reading, move the specimen traverse controls to make sure a grid bar is not shutting off the beam. Then centre the controls to make sure you have not simply moved off the specimen altogether. If the beam still does not appear, return all microscope controls to their normal idling positions and make sure nothing is in the path of the beam. Remove the objective aperture. If the beam appears, reinsert the aperture and centre it properly. If there is still no beam after these procedures, the routine alignment procedure may need to be repeated.

The above-mentioned causes of trouble plus a few others are summarized in Table 6 in the chapter on routine maintenance, pp. 190–192.

Emergency Shutdown

Should anything go wrong during operation of the microscope and you do not know how to handle the situation, turn the microscope off at once, following the 'emergency shutdown' rules in the instruction manual. A copy of these rules should be kept in plain sight so you do not have to hunt for them in time of emergency.

Most microscopes are provided with certain fail-safe devices which automatically turn off the high voltage when certain things go wrong. Situations in which such a turn-off will occur include a rise in pressure in the vacuum systems, a fall or increase in the pressure of the water cooling the lenses and diffusion pump, and instability in the high voltage supply caused by serious contamination in the region of the electron gun. Primary concerns when something serious goes wrong are to get the high voltage off (usually automatic) and the diffusion pump closed off from the column.

Isolation of the diffusion pump from the column is *extremely important* for the following reasons:

1. If the water which cools the diffusion pump jacket ceases to circulate, the oil vapour will not condense and will diffuse up into the column and contaminate it. The oil will reach such a high temperature that it 'cracks'. (Cracking breaks down the long-chain molecules of the oil, making it more volatile.)

2. If the mains electricity fails, the diffusion pump will continue pumping due to its thermal lag. The rotary pump backing the diffusion pump, however, will stop pumping. Unless there is a device fitted to the vacuum system which isolates the diffusion pump from the backing line when a mains electricity failure takes place, oil and vapour from the rotary pump will be drawn into the high vacuum system until equalization of pressure occurs. The oil from the rotary pump can be thermally decomposed by the diffusion pump and the vapour can enter the microscope column.

WHAT TO DO UPON COMPLETION OF WORK

When your work on the microscope is completed for the day, the beam should be defocused with the condenser 2 fine focus control, the filament current turned down and switched off, and the magnification controls returned to the lowest value. The specimen should be removed from the specimen chamber and stored in a safe place if you wish to keep it. If it is left in the microscope it will very likely be discarded by the next user. Gelatin capsules make handy containers for grids, or they may be placed on clean filter paper in a petri dish. If material on the grid has been worth photographing it is wise to keep the grid until the micrographs have been studied in case you wish to re-check something.

If someone else will be using the electron microscope that day, simply leave it with controls set in the proper idling positions described in the instruction manual. If it is not to be used again until the following day it can be closed down and left under vacuum, again following the routine in the instruction manual.

Electron microscopes are normally left under vacuum when not in use. This prevents the adsorption of air molecules on the walls of the column and speeds up the attainment of the optimum operating vaccum when preparing the machine for operation.

CONTAMINATION OF THE SPECIMEN

In the preceding discussion of typical operations involved in a day's electron microscopy, several references have been made to contamination and the problems which it causes. Even at a vacuum as high as 10^{-4} or 10^{-5} mmHg there are appreciable numbers of gas molecules within the electron microscope column. These fall largely into two groups: hydrocarbon molecules and water vapour molecules. The hydrocarbon molecules arise by evaporation from diffusion and rotary pump oil, grease used to help seal vacuum connections, rubber vacuum gaskets, parts which have been touched by the fingers before insertion into the microscope, and dirty specimen grids. These molecules will contaminate the specimen. Water vapour will evaporate from the photographic emulsion on the plates or film in the camera, even though these have been vacuum desiccated before insertion into the column. It can also arise from specimen holders and aperture rods which have not been kept dried.

When the electron beam is not on, hydrocarbon molecules will, from time to time, strike the specimen and at some variable time later move off again. Eventually an equilibrium is reached between hydrocarbon molecules moving on and off the specimen surfaces. When the specimen is being scanned with the electron beam, however, the electrons in the beam can interact with the hydrocarbon molecules resting upon the specimen surface in the illuminated region. First, polymerization of neighbouring hydrocarbon molecules occurs which hinders them from leaving the specimen surface. Then further bombardment of the polymer releases hydrogen from the hydrocarbon molecules, leaving behind a solid layer of pure carbon on the specimen. This not only obscures specimen detail but may also cause the section to bend, since the carbon layer and the specimen have different expansion coefficients. The rapidity with which this contaminating layer builds up depends upon a number of factors, including vacuum conditions, specimen temperature and column cleanliness. Obviously a scrupulously clean microscope column in which a minimum amount of grease has been used to help effect the vacuum seals will produce fewer contaminating hydrocarbon molecules.

Determining the Contamination Rate

The buildup of contamination on a specimen can be observed directly. One of the best types of specimen on which to observe this phenomenon is a carbon film with very small holes in it. At a magnification of say 60,000 to 80,000, allow the electron beam to fall at fairly high intensity on one of the holes. As contamination builds up, the edges of the hole will become covered with carbon and the diameter of the hole will decrease. The hole will finally be obliterated altogether. By photographing the hole twice on the same plate at an interval of 2 minutes and measuring the decrease in radius of the hole the contamination rate can be determined for any set of operating conditions. If the specimen is drifting slowly, this will also be obvious on the plate, as the second image of the hole will be displaced laterally. The contamination rate should not exceed 1 Å/s and should if possible be kept lower.

Removal of Material from the Specimen ('Stripping')

Water vapour molecules in the column can cause removal of material from the specimen itself. Water molecules on the specimen surface can be ionized by the electron beam. The oxygen thus released as a highly reactive free radical can oxidize carbon compounds in the specimen. The resulting CO, CO_2, H_2 and H_2O will then pass into the column and be evacuated. Some of the contaminating carbon molecules on the specimen surface are also removed in this manner, but not enough to prevent the buildup of this layer. To some extent the contaminating layer of carbon on the specimen surface actually protects the underlying specimen components from the action of the water ions, but it does so, of course, at the expense of good electron microscopic resolution. For a discussion of the effects of electron bombardment on specimen composition see Stenn and Bahr (1970).

Anticontamination Devices

Unless the electron microscope which you are using has an anticontamination device, the only way to avoid loss of resolution through contamination is to select your field rapidly, and focus and photograph as quickly as possible. In fact, to attain the ultimate resolution of which a modern electron microscope is capable, an anticontamination device is essential.

An anticontamination device is essentially a cold surface which completely surrounds the specimen except for the aperture through which the electron beam passes. The surface is cooled, usually by thermal contact with liquid nitrogen, and contaminating gas molecules are condensed on the cold surface before they can reach the specimen. At a temperature of $-80°C$ all hydrocarbon contamination ceases, but the specimen surface thus left exposed is subject to interaction with ionized water vapour molecules, and 'stripping' may occur. At a temperature of $-190°C$, however, the water molecules are inactivated to such an extent that the specimen can be viewed for an hour or more if necessary without observable change. A 'stripping rate' test may be performed to determine if the anticontaminator is performing adequately: the hole in a carbon film will get larger and begin to appear 'ragged' at the edge in about 5 min if stripping is occurring.

For a fuller discussion of contamination and its prevention, and for references to original papers on this subject, see Ruska (1966).

The Correction of Astigmatism

The nature of astigmatism was explained in Chapter 1. If the objective lens field is astigmatic the image cannot be brought into true focus, and image points appear to be elongated in one direction. If the condenser lens field is astigmatic the electron beam will, while being defocused at low magnification, appear elongate in cross section rather than round, and uniform illumination of the specimen becomes impossible at high magnifications.

To correct condenser astigmatism, a second field is introduced with the condenser astigmatism corrector. This causes elongation of the focused beam spot in a direction perpendicular to the elongation caused by the lens field asymmetry. By adjusting the strength and direction of the introduced field, the effects of the two fields can be made to cancel each other out and the cross section of the defocused beam becomes round. The correction should be made at a magnification not exceeding 1500X. The correction manipulations vary slightly from one microscope to another, but are easily accomplished with the aid of the instruction manual.

Correction of objective astigmatism is more difficult and requires considerable practice. It also involves introducing a compensating magnetic field to cancel out the effect of field asymmetry in the lens. Orientation of the field and adjustment of its strength with the objective astigmatism corrector requires very careful observation of the Fresnel (diffraction) fringes around some structure in the specimen. For this purpose a carbon film containing small holes is convenient to use. Fresnel fringes can be easily observed at the edges of the holes. Normally only the first maximum fringe will be visible and is henceforward referred to simply as the diffraction fringe, overfocus fringe, or underfocus fringe.

Using a perfect lens, free from all astigmatism, the edge of a hole in the collodion film would appear to have no Fresnel fringe associated with it when in true focus (Plate 3B). In reality, of course, a diffraction fringe *is* present due to the wave nature of the electron beam, but when the image is in focus the fringe is so narrow that it is not resolved by the microscope.

When the image is brought into the overfocused condition by increasing the current through the objective lens coil, the fringe increases in width and becomes visible as a bright line lying symmetrically *outside* the dark edge of the hole (Plate 3C). When the image is brought into the underfocused condition by decreasing the current to the objective lens, the fringe also increases in width, but is seen as a bright line lying *inside* the dark edge of the hole (Plate 3A). This underfocus fringe is extremely difficult to see without considerable practice.

Observation of the diffraction fringe associated with holes in carbon films can tell one a great deal about the condition of the microscope. The resolution being attained by the microscope can be determined by photographing the minimum overfocus fringe and measuring its width on the developed negative. The measurement must be taken on the fringe when the image is overfocused only sufficiently to just distinguish the fringe at, say, a magnification of 80,000 or more. The measurement is taken from the middle of the bright band to the middle of the dark band (which represents the edge of the hole) in order to determine the resolution (Plate 3C). This so-called 'minimum fringe' constitutes one of the most important test objects in electron microscopy.

If the objective lens is astigmatic, the holes in the carbon film cannot be brought into perfect focus. They will be overfocused in one direction and either focused or underfocused at right angles to this. The effect seen on the fluorescent screen is as depicted in

Plate 3D. The over- and underfocus fringes will be of minimum width when the object is nearest to being in focus. Any point observed under astigmatic conditions will appear to be elongated, as seen in Plate 3D.

To correct an astigmatic image such as seen in Plate 3D, the following steps should be performed:

1. Select a magnification higher than the magnification you plan to use for photographing.
2. With the astigmatism corrector turned off, determine the direction of the overfocus fringe on a hole which is small enough to be seen in its entirety.
3. Turn the astigmatism corrector on at full strength.
4. Adjust the direction of the astigmatism corrector field so that the overfocus fringe which it introduces lies at right angles to the direction of the overfocus fringe seen with the astigmatism corrector turned off.
5. Reduce the strength of the astigmatism corrector field until the fringes introduced by the two fields cancel each other out. A symmetrical fringe will now be seen around the hole. This fringe will disappear at true focus. The hole will now appear as in Plate 3B.

The astigmatism of the lens field has now been compensated. The compensation should, however, be checked frequently since it will not endure as contamination builds up on the objective aperture.

X-RAY HAZARDS

See Chapter 11, p. 180.

HIGH RESOLUTION ELECTRON MICROSCOPY

The foregoing sections have described operation of an electron microscope for *routine medium resolution work*. High resolution electron microscopy (resolutions of better than 25 Å for biological specimens) is an art in itself and is beyond the scope of this beginner's volume.

The requirements for high resolution work are exceedingly stringent. Special adjustment of the microscope is essential for success, as are extremely thin, high quality sections. Special criteria must be used in interpreting and assessing the results. For someone planning to do high resolution electron microscopy there is no substitute for an apprenticeship in a first-class electron microscopy laboratory where high resolution work is routinely done.

The techniques involved are exceedingly difficult if not impossible to absorb from written descriptions. However, some feeling for the methods required can be obtained from the excellent discussions of high resolution techniques to be found in the works listed below.

REFERENCES

Agar, A. W. (1965). In *Techniques for Electron Microscopy*, Chap. 1 (Edited by D. H. Kay), Oxford: Blackwell Scientific Publications.

Bahr, G. F. & Zeitler, E. H. (Eds.) (1965). *Quantitative Electron Microscopy. Laboratory Investigation* **14**, 1082–1158.

Haydon, G. B. (1968). On the interpretation of high resolution electron micrographs of macromolecules. *Journal of Ultrastructure Research* **25**, 349–361.

Meek, G. A. (1970). *Practical Electron Microscopy for Biologists*, Chap. 10 and 11 and pp. 400–402. London and New York: Wiley Interscience.

Ruska, E. (1966). In *Advances in Optical and Electron Microscopy*, Vol. 1, pp. 154–171, London and New York: Academic Press.

Siegel, B. M. (Ed.) (1964). *Modern Developments in Electron Microscopy*, pp. 51–78, New York and London: Academic Press.

Sjöstrand, F. S. (1967). *Electron Microscopy of Cells and Tissues*, Vol. 1, pp. 106–110, and Chap. 10 and 11. New York and London: Academic Press.

Sjöstrand, F. S. (1969). In *Physical Techniques in Biological Research*, Vol. 3C, 2nd edn, pp. 172–173.

Stenn, K. S. & Bahr, G. F. (1970). A study of mass loss and product formation after irradiation of some dry amino acids, peptides, polypeptides and proteins with an electron beam of low current density. *Journal of Histochemistry and Cytochemistry* **18**, 574–580.

6

Darkroom Procedures

If this is your first experience with the techniques of photography you will find this part of your training a most rewarding one. The fundamentals learned here will carry over both to photomicrography with the light microscope and to ordinary photography which you may wish to pursue as a hobby. Assuming that you have no prior knowledge of photography, the following information may be of value.

The Photographic Process

In conventional photography light waves reflected from the subject pass through an aperture in a light-tight box and impinge upon a light-sensitive emulsion which is supported either by a cellulosic or synthetic resin backing, as in roll film, or by a glass plate. The emulsion is basically a suspension of silver halide aggregates or 'grains' in gelatin. When light strikes a grain of silver halide, energy from the light is lost to the grain, resulting in the formation of small silver specks (latent image specks) which if large enough render the grain developable.

$$Ag^{+}Br^{-}(\text{solid}) \rightarrow Ag(\text{solid}) + Br$$

When the exposed film or plate is immersed in a developing solution in the dark, those silver halide grains which contain sufficiently large latent image specks are activated by the developer, which is essentially a reducing agent, and reduced to metallic silver. Thus a pattern appears upon the film or plate, forming a negative image of the subject which was photographed. The emulsion is then rinsed free of developer and immersed in a fixative which removes the opalescence and converts all the undeveloped silver halide in the emulsion to a water-soluble product which is removed by washing. The emulsion is now inert, and the dried negative can be stored until a positive print is required.

A positive print is produced by projecting light through the negative onto paper which bears a coating of photographic emulsion. Light passes through the negative in inverse proportion to the varying densities that make up the negative image. A latent image is not induced in the silver bromide grains in the paper below the opaque areas in the negative, whereas the lighter areas allow light to pass so that aggregates in the paper beneath them are energized and will darken upon development. After exposure, the photographic paper is developed, fixed, washed and dried as was the photographic film or plate which formed the negative.

The same methods are used when electron photomicrography is undertaken. The only difference is that an electron beam is used as an energizing source instead of a beam of light. Fortunately, the interaction of electrons with silver halide grains is sufficiently similar to that of light to permit the emulsions used for light photography to be used for electron photography. A short discussion of the difference in interaction of electrons and photons with halide grains will be found on p. 104.

Developers

These are essentially solutions of reducing agents, usually either metol and hydroquinone or phenidone-hydroquinone. A developer which has given consistently good results in our laboratory with the glass plates commonly used in electron microscopy is Ilford PQ Universal developer. This is also used for printing the micrographs from the developed plates. It comes as a stock solution which is diluted before use.

Exposure to air causes oxidation of the developer with consequent loss of effectiveness. When the developer becomes discoloured it should be discarded. For critical work it should be renewed daily for best results. Care should be taken not to get developer on the hands, as it will cause dermatitis in susceptible individuals. The application of barrier cream before entering the darkroom will help in this respect.

Fixatives

These are usually based on sodium thiosulphate ('hypo', $Na_2S_2O_3.5H_2O$), and the reaction involved is AgBr (solid) + $2S_2O_3^{2-}$ (aqueous)$\rightarrow$$Br^-$ (aqueous) + Ag $(S_2O_3)_2^{3-}$ (aqueous). Commercial fixatives vary considerably in the speed with which they work. A convenient fixative for use with plates and photographic papers is Hypam Rapid Fixative, which fixes the image within 30 seconds. This comes as a stock solution and is diluted with four volumes of tap water before use. A hardener such as Hypam Hardener is frequently added to the fixative used for plates to render the gelatin of the emulsion more firm, thereby reducing the risk of damage to the plate during handling.

Plates

Glass lantern-slide plates such as Ilford EM6 are in common use for electron microscopy. They have thin, blue-sensitive emulsions with a conveniently small grain size for high resolution work. They can be rapidly desiccated since glass does not absorb water. However, they break easily and consume a great deal of storage space.

Roll Film

The use of roll film in electron microscopy has the following advantages: A number of photographic exposures can be stored in far less space than can the same number of exposures recorded on glass plates. The cost per exposure is lower. Breakage is not a problem. However, its disadvantages have as yet discouraged widespread use. Roll film is easily scratched, and handling it in the darkroom is more difficult than handling plates. Desiccation takes longer. Its use is impractical unless the microscope is fitted with an exposure meter, and it is impossible to remove and process single exposures for study. Greater photographic enlargement is necessary than with plates and photographic 'grain' may become a problem.

Sheet Film

Sheet film is cut to the same size as a glass plate, and used in the same sort of camera magazine as are glass plates. It has the same advantages and disadvantages as roll film as far as storage, cost, scratching and handling are concerned. Some provision must also be made for keeping the film absolutely flat in its tray, since any buckling will change the magnification from one part of the sheet to another. One advantage over roll film is that single photographs may be taken, removed from the microscope and processed instead of waiting for the entire film to be used.

Appendix X lists photographic materials suitable for use in electron microscopy.

DEVELOPING AND PRINTING

The following procedures are carried out in the darkroom with the aid of an appropriate safelight (e.g. Kodak Wratten OB or Ilford S902).

The darkroom door should remain locked while you are working. Should you have to leave the darkroom before completing your work, be sure that all photographic paper and plates are in light-tight boxes before opening the door. Small amounts of paper may be kept separately for day-to-day use. Thus if a light leak does occur, loss of paper due to exposure is kept to a minimum.

The exact timing of the various steps will depend upon the type of developer and fixative used, the temperature of the developer, the type of photographic plate and the grades of photographic paper available. Unless there is someone available to tell you the optimum timings under the conditions in your particular darkroom it will be a matter of trial and error until you get what you want. Start by following the manufacturer's suggestions which accompany the various materials. Detail will be recorded on a plate or film by the right combination of exposure time and development time, the development time being determined by the manufacturer's instructions and the temperature.

The following routine is based upon the use of Ilford EM6 plates, PQ universal developer, Hypam rapid fixative, and Kodabromide or Ilford paper grades 1–4. Darkroom temperature is assumed to be 20°C (68°F).

Developing Plates

Properly exposed Ilford EM6 plates should come up to an optimum extent in PQ developer in 4 min. (The term 'come up' in photography refers to the gradual appearance of the image on the photographic paper as development proceeds.) If the electron microscope which you are using has an exposure meter, exposure presents no problem and development time can be standardized. If development time is standardized, however, the temperature of the developer should also be kept at a standard value. If there is no exposure meter, things are rather more complicated, since illumination on the fluorescent screen is difficult to judge by eye. A plate which has received too much illumination will come up too rapidly to ascertain when optimum development has been reached, and its background will appear 'muddy'. Too little illumination will result in a 'thin' (very pale) negative and loss of background detail.

During development the developer should be agitated. Ilford recommends constant agitation during development of EM6 plates. Associated Electrical Industries (makers of the AEI line of electron microscopes) recommend, in their notes for beginners in electron microscopy, 10 seconds agitation during each minute of development. Valentine (1966) stresses that agitation during the first part of development is particularly important to ensure that the image comes up evenly on the plate. Some workers bubble nitrogen gas through the developing tank. This serves the double purpose of agitating the liquid and removing oxygen which would otherwise oxidize the developer and shorten its life.

After development the plate should be rinsed for approximately 30 seconds to neutralize the developer before fixation, and to lengthen the life of the fixing bath. A few drops of glacial acetic acid is usually added to tap water for this purpose. This rinse is commonly referred to as the 'stop bath', since the acetic acid stops the action of the developer from proceeding further. Before passing the plates on into the fixative bath, the plates may be removed one by one from the stop bath and examined under a safelight to determine whether development has proceeded far enough. If one or more plates appear to be underdeveloped, they may be replaced in the developer after a quick rinse in tap water to remove the acetic acid. If care is taken, overdevelopment should not be a problem. If, however, a plate is overdeveloped this can sometimes be corrected by the careful use of Farmer's reducer (see Appendix X). This can be done at any time, so overexposed plates are usually fixed along with the properly exposed ones and dealt with at leisure another time. Serious overdevelopment cannot be corrected.

When you have decided that optimum development has occurred, the plates are transferred from the rinsing bath to Hypam rapid fixative. They remain in the fixative for approximately twice as long as it takes for the plate to 'clear'. ('Clearing' means that the opalescence on the plate is lost and the image becomes clearly visible.) Fixation occurs in two stages:

First stage. Opalescence disappears. Silver thiosulphate is formed which is insoluble and will cause staining if not solubilized.

Second stage. Insoluble silver thiosulphate is converted to a soluble complex salt.

Clearing (i.e. the first stage) should occur before 2 min have elapsed. If it does not, the fixing solution has been exhausted and the second stage will not occur. The fixative should therefore be replaced by a fresh solution.

Little harm will result if the plates are inadvertently left in the fixative for a longer time, although a bleaching of the plates may occur after long exposure to fixative.

Following fixation, the plates are rinsed for a minimum of 30 min in running tap water to remove the fixative and byproducts in emulsion and base, and then allowed to dry in air in a dust-free place. The plates may be exposed to light at any time after the first stage of fixing is completed.

PRINTING FROM THE PLATES

Kodabromide and Ilford papers are graded in series from 0 to 6. The numbers refer to the degree of contrast the paper is capable of providing. The higher the number, the higher the contrast. However, it should be remembered that the higher the contrast the less background detail will appear in the finished photograph.

A properly produced plate should have sufficient contrast and sufficient detail within it to come up well on grade 2 paper. Grades 0 and 1 are seldom used, as the contrast rendered is below the optimum for an electron micrograph. Grade 3 paper is used when the contrast produced by grade 2 paper is inadequate, and the background detail with grade 3 is usually fairly satisfactory. Grade 4 paper is used only on plates which have been seriously underexposed or underdeveloped. A plate requiring grade 5 or 6 had best be discarded, as background detail will be negligible. Usually, the paler and more washed out the appearance of the plate, the higher the grade of paper which must be used.

Prints from photographic plates are made by placing the plate into a holder in an enlarger. Light is projected through the negative onto photographic paper, which is held in place by a holder positioned below the enlarger. The enlarger is fitted with a diaphragm which can be 'stopped down' to reduce the amount of light passing through the lens. A 'thin' (i.e. underexposed) plate allows more light to pass the lens than a properly produced plate, and this can be partly compensated for by 'stopping down' the diaphragm. In general, longer exposures at reduced illumination produce a photograph with more detail and better contrast than do short exposures at higher illumination. Short exposures (under 5 seconds) do not allow optimum recording of fine detail, and also do not allow time for shading (see paragraph below on how to deal with an unevenly exposed plate).

Exposure time varies inversely with the amount of light passing through the enlarger lens. A well-developed plate on grade 2 paper will usually require at least 6 seconds exposure at reduced illumination. The paper is then placed in developer. If too great an exposure is given, the image will come up too rapidly in the developer, the print will be too dark, appear 'muddy' and lack detail. Contrast will be poor. If an image appears on the paper before 30 seconds have elapsed the resulting print will probably be inferior. Prints which start to appear at about 45 seconds can usually be left for 2 to 3 minutes without overdevelopment. Correct exposure and development will produce optimum detail and contrast. Prolonged development will promote chemical fog and staining. It is best to standardize development time and temperature, and have only exposure variable.

When optimum development has been attained, the paper is rinsed in tap water containing a few drops of glacial acetic acid and fixed for 30 seconds in rapid fixative without hardener. Removal of the paper at precisely 30 seconds is not necessary. However, the raison d'être of the rapid fixative is that it will perform its job speedily, so this efficiency may as well be made use of. The paper is then rinsed in running tap water for a minimum of 30 min. Inadequate rinsing may result in darkening or discoloration of the print within a short time. Longer rinsing is unnecessary and may eventually cause the photographic paper to become limp.

After rinsing, the prints are placed on a print drier. If a matt finish is desired the print is placed so that the image side dries in contact with the canvas cover. If a glossy print is desired (e.g. for publication) the print is placed so that the image side contacts the highly polished drum or a special glazing plate.

Working Prints

Obviously not all electron micrographs are potentially publishable, and some plates may not be worth printing. A badly out-of-focus plate will convey no useful information and should be discarded. Plates in or near focus, however, should be printed, even though a quick survey of the plate proves disappointing. Much detail in an electron micrograph is obvious only after careful study of a print, and much information may be gleaned from a micrograph which is not aesthetically pleasing. One might not publish a micrograph marred by stain precipitate or by a knife score, but it might provide an invaluable bit of confirmatory information which could be incorporated into written results.

Usually 'working prints' taken from electron micrograph plates are printed at the same size as the plate or slightly larger. This allows the use of small-sized paper, and the micrographs can then be filed and stored in loose-leaf folders. It is usually not worth while wasting paper in trying to get optimum results in these working prints. As long as they are of sufficiently good quality to show up the detail you need they can be used for study. When particular micrographs have been selected for publication or display, then time and paper may be expended on getting the best possible results.

Enlarging Micrographs

If a plate is in focus, an enlargement of six or seven times will make visible to the naked eye all of the detail inherent in the photographic plate. This is because the eye can see two points as separate entities if they are 100 to 200 μm apart; i.e. the resolution limit of the eye is 100 to 200 μm. An enlargement of the plate six or seven times will result in the 'grain' of the emulsion becoming resolvable by the eye. A further enlargement may aid visualization of the fine structure, but no more meaningful detail will be resolved.* For relatively low resolution studies an enlargement exceeding six times is seldom of practical use.

Contrast decreases as enlargement increases, so a higher grade of paper will be needed to compensate for this.

At times a plate which appears to be in focus in the working print will be obviously out of focus when enlarged. Such a plate is, of course, not worth printing at a higher magnification. A little practice will enable you to detect such an out-of-focus plate on the enlarger before it has been printed, thus saving paper.

Photographic paper is very costly, and it is seldom necessary for the beginner to enlarge a print over the usual working print size. Enlargement should be undertaken only if measurements are to be made, or if an important structure has been photographed the details of which cannot be clearly seen at the magnification of the working print.

How to Deal With an Unevenly Exposed Plate

If the photographic plate was not evenly illuminated when the micrograph was taken, the prints taken from it will have areas which come up at different speeds during development. If the

* It should be noted that 'grain' in this sense does *not* refer to the size of the silver halide aggregates themselves. In the emulsions used for electron microscopy these are approximately 0·5 to 1 μm in diameter and thus a magnification of 100 to 200 times would be necessary for them to become visible to the naked eye. The 'grain' seen in the plate enlarged six or seven times is caused by 'electron noise'—the interaction of two random processes occurring during exposure. For a discussion of these phenomena see Valentine (1966).

unevenness is not too great it can be compensated for in some cases by 'shading' the print. This is done by passing one's hand rapidly back and forth during exposure of the photographic paper so that the moving shadow of the hand reduces the amount of light striking the paper in a particular area, care being taken not to position the hand too near the paper. For example, if the left end of the plate is darker than the right, expose for a length of time which will be adequate for the darker end. During the exposure, keep the shadow of your hand moving across the photographic paper in such a way that the amount of light reaching the paper from the right (paler) side of the plate is cut down. Be sure to keep the hand moving, or you will simply produce an image of your hand on the paper!

Perhaps, due to improper spreading of the beam at time of photography, the plate is too dark at the corners but well exposed in the middle. Unless there is important detail at the periphery, probably the simplest thing to do is to enlarge the micrograph until only the evenly illuminated central portion is included in the print.

A Note About Darkroom Cleanliness

While working under safelight conditions some slopping of developers, rinsing baths or fixatives is likely to occur. These spills look like water. However, the fixative will dry down into a white powder and the developer into a multicoloured sludge. When dry these tend to blow about, leaving a layer of contamination on everything. Therefore, when you are finished in the darkroom it is good practice to wipe down the benches with a clean sponge, rinse the sponge and wipe down again. Be sure that developing and fixing solutions are poured back into the appropriate bottles if they are still usable, or discarded if they are not. Rinse the pans thoroughly, dry them and put them away.

The Difference Between the Action of Electrons and Light on Photographic Emulsions

When a photographic emulsion is exposed to light, it takes a number of photons to produce a latent image in a silver halide grain. When a photographic emulsion is exposed to electrons, however, a single electron has sufficient energy, at the voltages employed in electron microscopy, to cause a latent image to form in a silver halide grain. The interaction between electrons and halide grains is therefore described as a 'single hit process'. This has practical consequences as far as the relation of contrast to exposure time is concerned.

The density of a plate or film is spoken of in terms of a scale of 0 to 3. A plate of 0 density would be transparent when developed since no exposure to light or electrons has occurred, no latent images have been formed, and therefore no development takes place. Transmission of light through such a plate would theoretically be 100 per cent. A plate where $D=1$ when developed allows 10 per cent transmission of light, and appears light grey. $D=2$ permits only 1 per cent transmission and is quite dark. $D=3$ permits only 0·1 per cent transmission and is practically opaque. Electron microscopists usually endeavour to attain plates of $D=1$ to 1·5.

The density of a plate is linearly related to the number of developed silver grains. The density is not, however, linearly related to *exposure time* in the case of plates exposed to light. This is because of the multiple-hit nature of the interaction of photon with halide grains. The larger the grain of the emulsion the more photons will hit a grain and the faster a latent image will develop within the grain. As a consequence, the contrast obtained with a given exposure time to light is *dependent upon the particular emulsion used*, as well as on exposure time and conditions of development.

With plates exposed to electrons, however, density is *linearly* related to exposure time over the range of exposure time used in electron microscopy (1 to 4 seconds). Because one hit by an electron develops one grain, no matter what the size of the grain, the contrast obtained on a plate is *independent of the particular emulsion used* and is directly proportional to exposure time. Contrast depends only on exposure time and conditions of development.

The prime determinants in choosing an emulsion for electron microscopy are the grain size, speed, emulsion thickness and relative light sensitivity.

Grain size in emulsions used for electron microscopy should be small (0·1 to 1 μm) so that the developed silver grains will delineate fine detail. A slow emulsion is desirable in order to get a large number of electrons onto the plate. However, a slow plate requires a long exposure, and after 4 seconds instabilities in the microscope become a limiting factor. Therefore a compromise is made and a medium speed emulsion selected. Penetration of electrons increases with accelerating voltage. Emulsion thickness should be sufficient to absorb all the electrons in the beam completely. Twenty-five micrometres thickness is sufficient in the 20 to 100 kV range.

It is convenient to work with blue sensitive emulsions which are least sensitive to light and which can therefore be processed under an appropriate (orange) safelight.

For a full discussion and mathematical treatment of the relationships between exposure time, density and contrast, see Valentine (1966).

Enlargers

There are basically three types of enlarger available:

1. *The Cold Cathode Enlarger*
Light is produced by several fluorescent tubes and is then passed through a diffusing screen in order to illuminate the plate evenly. The diffusion reduces contrast and limits the amount of fine detail producible on a print. This type of enlarger is *not* recommended for electron microscopy.

2. *Opal Lamp and Condenser Enlarger*
Light is produced from a coiled filament and is semi-diffuse. It is less expensive than a point source enlarger and is adequate for medium resolution work.

3. *Point Source and Condenser Enlarger*
The light is produced from the tip of a V-shaped filament. It yields a high-contrast print in which every detail is revealed. It also reveals every defect on the plate, and focusing is more critical. A point-source enlarger is essential for high resolution work.

Print Processing Machines

These cut down the time involved in printing. The plate or film is placed in the enlarger, and the photographic exposure is made on a special paper. The paper contains a developing agent in inactive form. The exposed paper is fed into the print processing machine where a chemical activitates the developer, another chemical fixes the image, and rinsing and partial drying are accomplished. The whole process takes about 10 seconds.

The main advantages of the print processing machine is the enormous saving of time. It has several disadvantages, however. Unless the plates or film being printed from have been correctly exposed a suitable print will not be obtained. Therefore a test exposure may be necessary for each plate, and the special paper is expensive. The prints will fade and discolour after a few months unless they are re-fixed and washed. This can, however, usually be done at any time within 6 months of printing and a large number can then be done at one go. If the prints are to be dry mounted, they must be fixed and washed or they will stain when heat is applied. Also, the prints tend to curl and crack even when stored flat.

In general it seems to be felt that the advantages of print processers outweigh their disadvantages, especially in laboratories where a large amount of printing is routinely done.

Kodak and Ilford both supply print processing machines (see Appendix, p. 223).

REFERENCES

Engel, C. E. (Ed.) (1968). *Photography for the Scientist*, London: Academic Press.

Farnell, G. C., Saunders, A. E. & Flint, R. B. (1970). The response of photographic materials to exposure in the electron microscope. *British Journal of Photography* **117**, 139–141.

Meek, G. A. (1970). *Practical Electron Microscopy*, Chap. 13, pp. 343–360, London and New York: Wiley Interscience.

Sjöstrand, F. S. (1967). *Electron Microscopy of Cells and Tissues*, Vol. 1, Chap. 5, London and New York: Academic Press.

Valentine, R. C. (1966). In *Advances in Optical and Electron Microscopy*, Vol. 1, pp. 180–203 (Edited by R. Barer and V. E. Cosslett), London and New York: Academic Press.

7

Some General Suggestions for the Study of Ultrastructure

The term ultrastructure is used in this book to refer to structure which lies beyond the resolving power of the light microscope. One of the principal tools for its study is the electron microscope. Biochemistry is another.

From the beginning, electron microscopy has tended to fall into two distinct categories: low-to-medium resolution electron microscopy on the one hand and high resolution electron microscopy on the other. The linear dimensions at which medium resolution leaves off and high resolution begins have decreased with time, as improvements in electron microscopes and in specimen preparative techniques have been introduced. It still depends upon the type of material being studied. A resolution near 2 Å has been achieved with crystalline specimens. Reports claiming a resolution of 10 Å or less on ultrathin sections of biological material are very few indeed (e.g. Sjöstrand, 1963, 1969; Haydon, 1969*a*), and 15 Å is usually considered the practical limit.

High resolution electron microscopy will be arbitrarily considered in this book to deal with structure in biological materials which is revealed only with a resolution of better than 25 Å.

Low-to-medium resolution electron microscopy has permitted the morphological description of cell organelles, and through cytochemical techniques has provided some general information on their chemical nature. It extends light microscopy and, since it can be used to examine the morphology of isolated cellular fractions, it overlaps with and complements biochemistry.

High resolution electron microscopy is concerned with the study of the molecular and macromolecular structure of protoplasm. It constitutes a division of molecular biology. It involves techniques which are far more demanding than those of low and medium resolution electron microscopy. The interpretation of its results presents problems which do not confront electron microscopists who work at a lower resolution. Among these problems is the interpretation of phase contrast and diffraction effects on the image. Slight defocusing increases contrast by introducing a minimum diffraction fringe around specimen structures. Greater defocusing will produce a so-called 'defocus granularity' which is seen over the entire field even when no specimen is present. This is easily confused with molecular structure (Haydon, 1968; Sjöstrand, 1969). Also, the exact effects of the fixatives, stains and other processing materials on the molecular configuration of entities seen at such resolutions is unknown. Studies by Lenard and Singer (1968) and Sjöstrand and Barajas (1968) have indicated that protein structure is drastically altered by the fixatives commonly used in electron microscopy. Effects of the electron beam upon molecular configuration must also be considered.

The present volume, as a beginners' handbook, does not discuss in detail the techniques and problems of high resolution electron microscopy. The reader should only note that a dividing line does exist between the low-to-medium resolution range and the high resolution range. References at the end of the chapter should be consulted regarding high resolution work. For those persons planning to do high resolution work, 6 months training in the laboratory of an experienced high resolution electron microscopist would be invaluable.

The present chapter deals with the interpretation of electron micrographs which are the result of techniques which will probably only allow a resolution of about 25 to 50 Å. This approximate limit is set by a number of variables. Some of the variables concern the electron microscope itself and lie outside the control of the operator. These include the focal length and numerical aperture of the objective lens, and the voltage settings provided. Other variables concern preparative techniques, one of the most important of which is section thickness. With some microscopes the sections must not exceed a thickness of 150 Å to attain a resolution of 20 Å. Other microscopes will give the same resolution on a section ten times as thick, as long as certain technical details are carefully controlled (Wachtel, Gettner and Ornstein, 1966).

THE STUDY OF ULTRASTRUCTURE: PRELIMINARY CONSIDERATIONS

Studying Published Micrographs

It is assumed that anyone wishing to use the electron microscope to study biological tissue has had a grounding in histology and basic cytology. This will afford a framework into which the information contained in electron micrographs may be fitted. Anyone who has not had a recent course in cytology should read one of the cytology textbooks listed in the references at the end of this chapter.

If you have not previously studied electron micrographs you are advised to do so carefully for a short time each day for several weeks. In this way, while you are learning to run the electron microscope and to prepare tissue for study you can also learn what to look for. A feeling for ultrastructure cannot be developed in one all-night session. Do not confine yourself to the one type of tissue that you happen to be interested in. Certain structures are common to all cell types, although they may vary in very subtle or very dramatic ways from tissue to tissue.

Several excellent atlases of ultrastructure which can be used for this purpose are listed at the end of the chapter. Begin with Toner and Carr (1966) which gives a general background. Then go through Porter and Bonneville (1968) which deals with ultrastructure in the context of the various tissues of the body. Then go on to Fawcett (1967) which deals in more depth with the various organelles and their fine structural variations. Carefully study each micrograph and check it against the figure legend. This will get you used to the sort of structures which are seen in electron micrographs and how their appearances differ with plane of section. The more sorts of structure you become familiar with, the easier it will be to identify things in your own micrographs and the less likely you will be to miss something important, or to submit for publication a picture of a variant of some well-known cellular component under the heading 'New Cytoplasmic Structure Discovered'. It is also useful to go through either Ham's *Histology* or Bloom and Fawcett's *A Textbook of Histology* (latest editions, of course) and compare their electron micrographs of the various types of tissue with their light micrographs. Always consider ultrastructure in the light of function.

Background Work With the Light Microscope

Before beginning work on a particular tissue at the ultrastructural level, a thorough knowledge of its architecture at the level of light microscopy is essential. Some of the necessary background can be obtained from textbooks of histology. Most texts, however, describe human histology, so species differences must be considered if work is to be done on another animal. The value of studying a set of serial sections of the organ to be investigated is immense. If this is not feasible, enough sections should be studied so one can determine later from 1 μm sections taken from the tissue before ultrathin sectioning approximately what one has incorporated into the 1 mm block and whether, after one has sampled several 1 mm blocks from a single organ, one has obtained a reasonable sample of all the cell types to be found within it.

A far more detailed and meaningful study with the light microscope is possible with 1 μm sections of tissue which has been prepared for electron microscopy than is possible with 5 μm sections of paraffin-embedded tissue. This is partly due to the excellence of glutaraldehyde fixation, with or without post-osmication, and partly due to the thinness of the section. Detail of cellular structures in 1 μm sections is far superior to that seen in ordinary histological preparations. Cell boundaries are distinct, cytoplasmic organelles suffer little or no displacement or disruption, and extraction and distortion of nuclear and cytoplasmic matrix is minimal compared with other routine methods. There are a number of staining techniques which can be applied to 1 μm sections of epoxy-embedded material, a few of which are reported to have a degree of specificity. Routines for some of these are given in the Appendix.

The Importance of Orientation Slides

It has been estimated that to cut serial ultrathin sections for electron microscopy through a block of tissue one cubic millimetre in volume and to photograph the entire area of each section, would take approximately 4×10^8 photographic exposures.

If a 1 mm^2 section is taken from the face of such a block, an area of one million square micrometres is produced, and in a closely packed tissue with cells averaging 15 μm in diameter, profiles of over 4000 cells could be present in a single section. The amount of tissue obscured by the metal of the supporting grid varies with the size and shape of the mesh, and may approach 65 per cent in grids with small circular holes. Nevertheless, detailed electron microscopy of the visible tissue in such a section would entail many hours of work. Even the location of specific details by quick scanning would be time consuming if not impossible, due to the discontinuities produced by the supporting grid.

The time spent scanning with the electron microscope is most easily minimized if the specific area which one wishes to study is first located by light microscopy, as described in the chapter on sectioning. The 1 μm sections prepared for orientation purposes can also be of invaluable assistance in interpreting the electron micrographs of ultrathin sections. The 1 μm sections can be made into semi-permanent preparations by drying the slide on a warming tray or in the oven and then applying a coverglass with some suitable medium such as DPX. Some stains such as methylene blue will fade if exposed to light, but will remain useful for many months or even years if kept in the dark. It is said that use of epoxy glue as a mounting medium will retard fading (Harrison, 1971).

Of course the 1 μm sections will contain more tissue than has been incorporated in the ultrathin sections taken from the block after trimming. Therefore, after the block is trimmed, it is advisable to make a rough diagram of the appearance of the 1 μm section as seen in the light microscope and indicate upon it the portion which has been left on the face of the trimmed block. This sketch may be filed with the orientation slide. Then prior to scanning the ultrathin sections made from the block, the orientation slide and diagram can be retrieved for study. This will facilitate recognition of the features which will be seen with the electron microscope.

If the grids prepared at each cutting are placed in individual containers and numbered consecutively, and then a second orientation slide made after the last ultrathin section has been taken, a fairly accurate idea of what should appear on the grids can be obtained.

It should be appreciated that if the block is properly aligned in the ultramicrotome so that little or no tissue is lost before a usable ultrathin section is obtained, and if cutting proceeds without incident, the picture presented by the second orientation slide may differ very little from that seen in the first orientation slide. If five grids are prepared, each with a ribbon of five sections upon it, and if the sections are of an average thickness of 500 Å, then:

$$5 \times 5 \times 500 = 12{,}500 \text{ Å, or } 1{\cdot}2\ \mu\text{m}$$

This represents only 8 per cent of the thickness of a cell with a diameter of 15 μm.

It should also be noted that the five sections on a single grid are serial to one another and represent a total thickness of only 2500 Å or 0·25 μm. Therefore if a structure within a cell is seen five times on one grid, the microscopist must take care to determine whether these five objects are occurring in five different cells, or whether they simply represent serial sections through a single structure in one cell.

Meaningful Samples

This leads us to a consideration of what a meaningful sample of tissue is in electron microscopical terms. Obviously, conclusions cannot be drawn from micrographs taken of tissue from a single animal, regardless of how many sections have been cut. The same criteria must apply here as in light microscopy: tissue from a sufficient number of animals must be sampled to avoid some freak abnormality being mistaken for the norm. The absolute number of animals varies, of course, with the type of study but four to six animals would appear to be a minimum in most cases. The age of the animals in each group should be approximately the same, as the ultrastructure of most organs varies from youth to senescence. How much it varies, in fact, has not been fully determined. The state of nutrition, weight, sex, and (if the animal is a female) how many litters she has produced, can all be important factors in some work.

If the tissue to be studied is a complex one, it is particularly difficult to sample the tissue properly. One cannot, for example, just cut up an adrenal gland into 1 mm cubes, shake them about, process them, pick out a few cubes at random and expect to learn much about the structure of the adrenal gland at either the microscopic or ultramicroscopic level. The cortex of the adrenal gland has three zones, each of which differs from the others both in obvious and in very subtle ways. A cubic millimetre of tissue from the glomerulosa will tell you nothing about the reticularis. The adrenal medulla is an altogether different sort of tissue from the cortex. It has a separate embryological derivation and a function entirely dissimilar to that of the cortex. In studying an organ such as the adrenal, great care must be taken to ensure that you obtain tissue from the particular zone you are interested in, and that samples have been obtained from several areas within this zone. It is helpful to fix thin slices no more than 1 mm thick, rather than 1 mm cubes, and embed the slices flat. In this way one gets a larger area for study with a light microscope for orientation purposes.

Even a tissue which is fairly homogeneous, such as the liver, requires careful attention to sampling. Although the epithelial cells (hepatocytes) are considered to be a single cell type, they perform a great variety of functions. Cell ultrastructure may vary with function; function may vary with the position of the cell in the liver lobule.

THE STUDY OF ULTRASTRUCTURE: EXAMINATION OF ONE'S OWN MICROGRAPHS

It is essential to make yourself familiar with a particular tissue by means of the light microscope before studying it with the electron microscope. The importance of knowing in great detail what has been learned about that tissue by other electron microscopists is not as essential at this stage. In fact, perversely, it may be a handicap. Too strict attention to what one knows should be there can produce a state of mind in which one sees only what one expects to. Probably the most effective way for the novice in electron microscopy to gain experience is by looking at the tissue —a lot of tissue—with a fresh eye, and describing what he sees *in writing*. This is a highly personal view, and the student may find that with time and experience he develops an entirely different approach. However, this method is suggested as a good way to start.

In studying a set of micrographs taken of the grids obtained at a single cutting it is helpful to get out the orientation slides pertinent to the cutting and describe the area included therein in detail. It is surprising how the act of putting what one sees into words makes one see more.

Do the same with the electron micrographs. Describe as much of the general architecture of the tissue as is visible in the micrograph. Describe the cells. Are they all one type? How do you know? How are the cells related to one another? Are there spaces between them or are they tightly joined together? If there are spaces are they artificial? Are there desmosomes or other forms of intercellular attachment? Is there any material visible on the free surfaces of the cell membrane? Are there any indentations of the cell membrane? Any pinocytotic or other vesicles just beneath it? If so, describe their contents. Are the cell membranes intact? Are there signs of swelling or breakage from fixation artefacts or rough handling?

Is the cytoplasm of the cells homogeneous or patchy? Pale or fairly electron dense? Are there any fine filaments in it? Are these really filaments, or sheets of material cut edge on? How do they appear when sectioned from another angle? Are their cross sections round?

What do the mitochondria look like? Are they confined to one part of the cell section, or widely distributed? Are they round, filamentous, branched, single or in clusters? Are they swollen? Is the mitochondrial matrix a uniform density? Are any granules present in the matrix? In what form are the cristae?

Is rough or smooth endoplasmic reticulum present? Is it confined to a particular area of the cytoplasm? Is it in stacks or single cisternae? Are all the ribosomes bound to the endoplasmic reticulum, or are some of them scattered through the cytoplasm? Are the scattered ribosomes distributed singly or in clusters?

Are there any cytoplasmic bodies present which you cannot identify? Are there any areas peculiarly free from organelles where some substance may have been extracted?

Is the Golgi apparatus in evidence? Does it have both cisternae and vesicles? What size distribution of the vesicular component is seen? Does any electron-dense material appear in any of the cisternae or vesicles? Are one or more centrioles associated with the apparatus?

Are the nuclei regular in outline? If irregular, does this indicate a shrinkage artefact or reflect a particular state of physiological activity within the cell? Are there indentations of cytoplasm into the nucleus? Is there any indication that material might be passing between nucleus and cytoplasm? Is a nucleolus present? What form does it take? How many of its components are in evidence? Does the nucleolus lie near the centre of the nucleus, or is it associated with the nuclear membrane?

After a number of cuts through various specimens of the tissue under study have been described, one has gained a 'feeling' for the tissue—a familiarity not easily obtained in any other way. Then it is time to go to the literature and read intensively. If the tissue has received considerable attention from others you will find many of the things you have described mentioned in the literature. Perhaps you have drawn a tentative conclusion about a particular organelle and someone else has described the same thing but drawn an entirely different conclusion. Perhaps *you* are right—think about it and read on.

Remember species differences. A rat is not a man. If you have chosen to work on iguanas you probably have the field to yourself.

ARTEFACTS

One of my students, taking his degree examination, defined an electron micrograph as a photograph of the shadow of an artefact. (He claims the definition is his own.) When one considers the amount of material which has been removed from the cell during processing one has to admit that this statement is not unjust. About 60 per cent of most cells is water. This is removed during dehydration in the alcohols, and is replaced first by the dehydrants and then by the embedding medium. Of the remaining 40 per cent, a variable proportion of the proteins, lipids, carbohydrates and nucleic acids are removed either during fixation or later during dehydration and embedding. What remains after fixation in glutaraldehyde followed by osmium tetroxide is a cross-linked skeleton composed of those materials (mostly structural protein and molecules bound thereto) which the fixative has 'stabilized'. Considering that probably only 20 to 30 per cent of the original material of the cell remains, it appears quite remarkable that the image seen in the electron microscope has any relation to reality at all.

Nevertheless, by comparing the results obtained by the use of different fixatives and different dehydration media, and also the results obtained by the methods of freeze drying, freeze substitution, and freeze etching (see Chapter 10), a picture of cellular ultrastructure has emerged which is remarkably similar from one method to another.

There are, however, certain artefacts which are commonly seen in micrographs, and these the beginner should learn to recognize. The ones discussed below are recognizable in the medium resolution range. In this range one is dealing with what might be called the 'gross anatomy' of the cell—the appearance of its nucleus, mitochondria, lysosomes, and membranes as seen at resolutions of 25 to 50 Å. At high resolutions (below 25 Å) one reaches the macromolecular level where another spectrum of artefacts becomes important.

Causes of artefacts tend to fall roughly into three main groups: (1) mechanical damage due to rough handling, (2) chemical, physical and osmotic disturbances and extraction of materials incurred during fixation, dehydration and embedding (processing artefacts), and (3) artefacts introduced after the tissue has been embedded, e.g. by sectioning damage, dirty working conditions, stain precipitate and damage by the electron beam. Those in groups one and three are usually readily identifiable as soon as the specimen is viewed. Those in group two may be easily seen or hard to detect —in either case assigning their origin to a particular stage in the processing routine is extremely difficult. Some of the symptoms indicative of the three groups are listed below.

1. *Mechanical Damage*

This may manifest itself as torn cell membranes, ruptured cells, and cell organelles or red blood corpuscles found in intercellular spaces. Lesser degrees of damage may be difficult to differentiate from processing damage.

2. *Processing Artefacts*

Severe processing damage will be obvious as soon as a 1 μm section is cut from a block. The inner part of the tissue may crumble away, or the staining properties of the sectioned tissue may change from the periphery to the interior. If the 1 μm section appears to be adequately preserved, processing artefacts

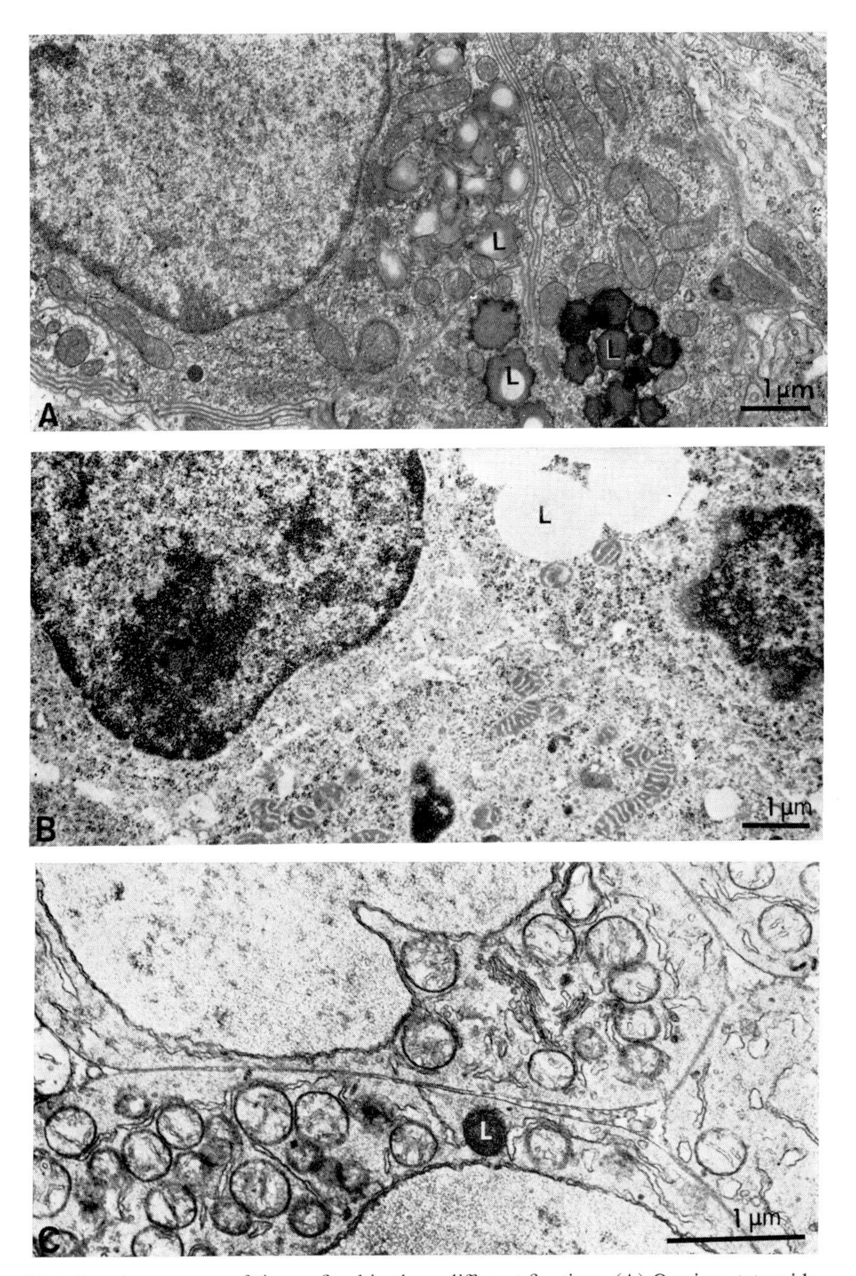

Plate 1. Appearance of tissues fixed in three different fixatives. (A) Osmium tetroxide. Membranes are clearly seen, lipid droplets (L) are preserved to a varying degree, ribosomes are present, nucleus is fairly homogeneous. (B) Glutaraldehyde. Membranes are seen as negative images, lipid (L) has been extracted leaving white areas, ribosomes are present, nuclei show more clumping of chromatin than in (A). (C) Potassium permanganate. Membranes are clearly seen, lipid (L) is present, cytoplasm and mitochondrial matrix are pale due to leaching out of protein, ribosomes are missing, nuclei show no chromatin.

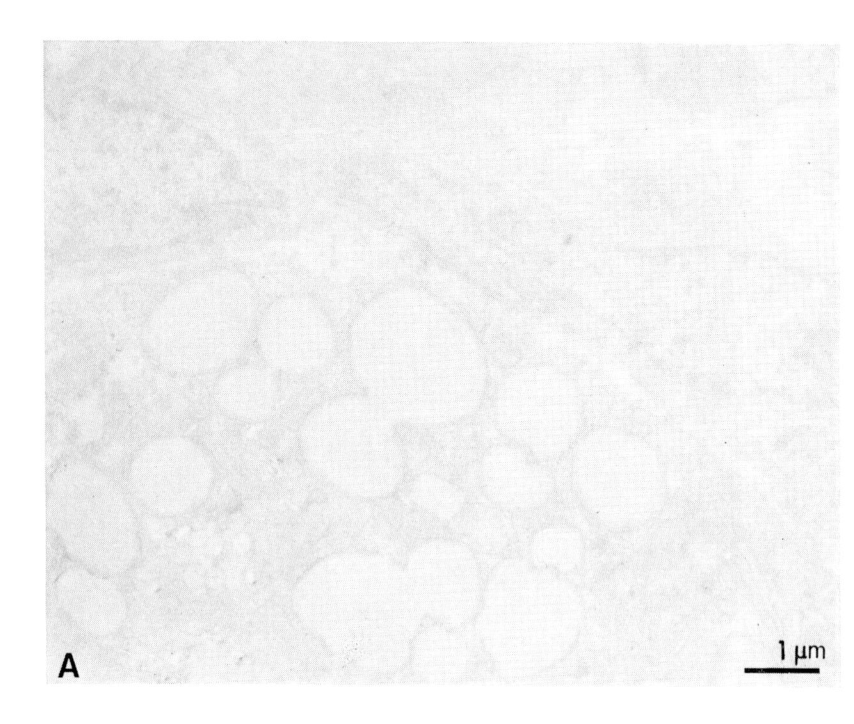

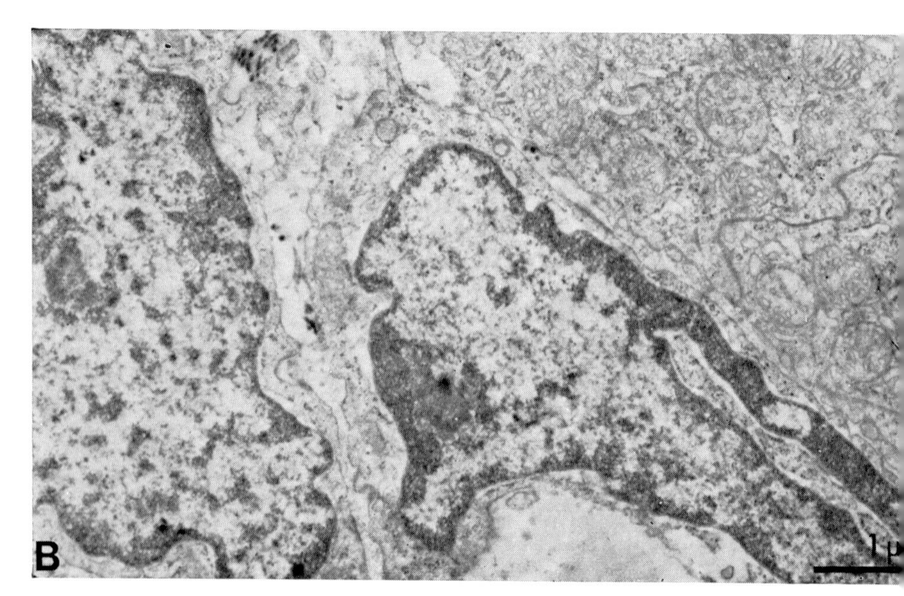

Plate. 2 Effect of uranyl acetate and lead citrate stains used alone and in combination. The tissue has been fixed in osmium tetroxide followed by glutaraldehyde. (A) Unstained tissue. Contrast is very low. (B) Uranyl acetate (2 per cent aqueous). There is a general increase in contrast. The most intense staining is seen in the nuclear chromatin and cytoplasmic ribosomes.

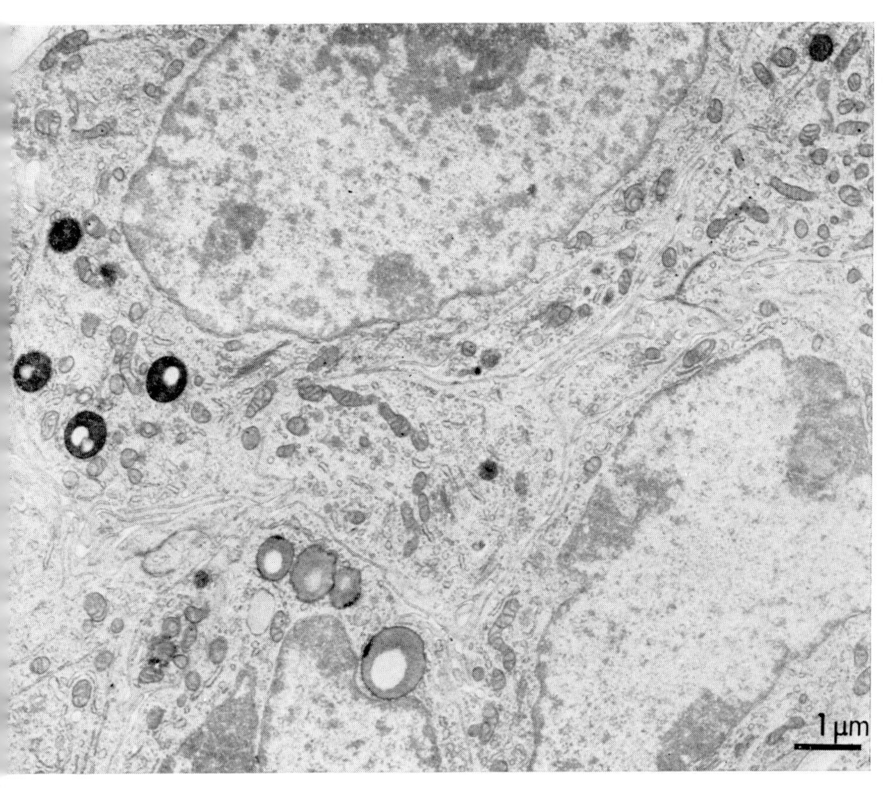

Plate 2C. Lead citrate (Reynolds). There is an over-all increase in contrast. Membranes are particularly crisp and well delineated.

See over for plate 2D.

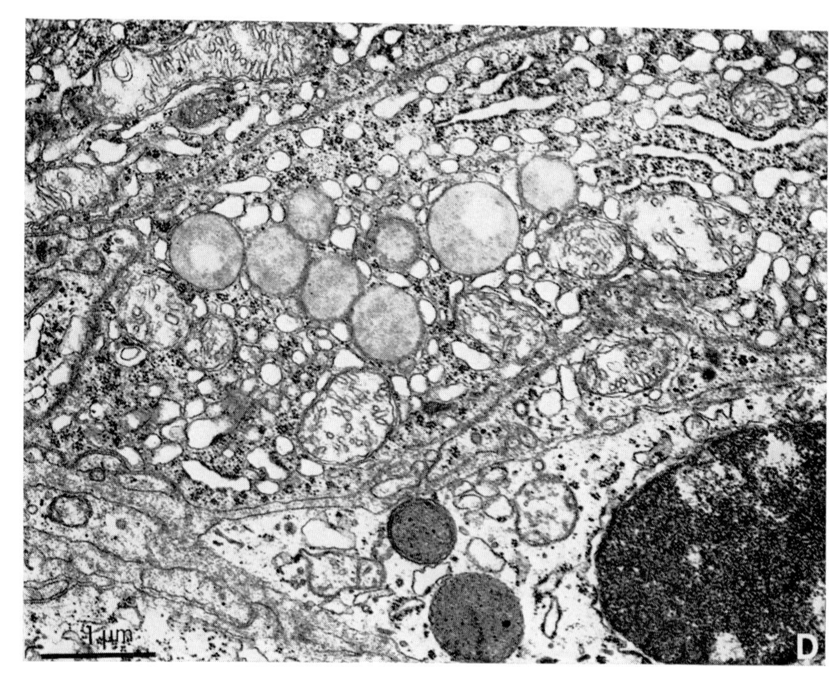

Plate 2D. Uranyl acetate (2 per cent aqueous) followed by Reynolds' lead citrate. The over-all contrast is greater than that with either stain used alone. Staining of the chromatin in nucleus at lower right is particularly intense; electron density of cytoplasmic ribosomes is very high.

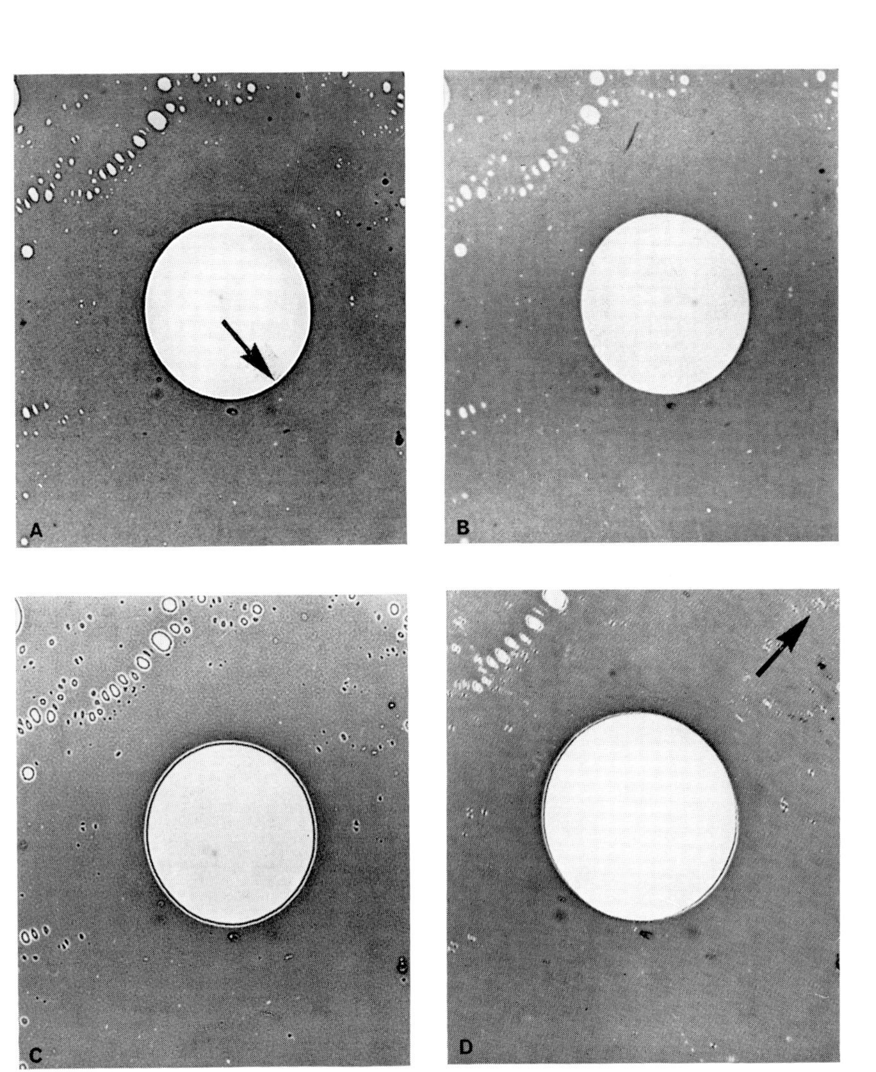

Plate 3. Through-focus series of photographs of holes in a carbon film. (A) Underfocused. A white line (the underfocus fringe) can be seen just *inside* the dark edge of the large hole (arrow). The carbon film appears granular (defocus granularity). (B) In focus. No fringe is apparent either inside or outside the holes. The carbon film has lost the granular appearance, contrast is at a minimum. (C) Overfocused. A white line (the overfocus fringe) appears just *outside* the dark edge of the holes. Defocus granularity is again apparent on the carbon film. (D) Astigmatism of the objective lens. The holes are in focus in one direction but over-focused in the other. This causes the defocus granularity of the carbonfilm to assume a directional character: i.e. the granules are elongated in the direction of the arrow.

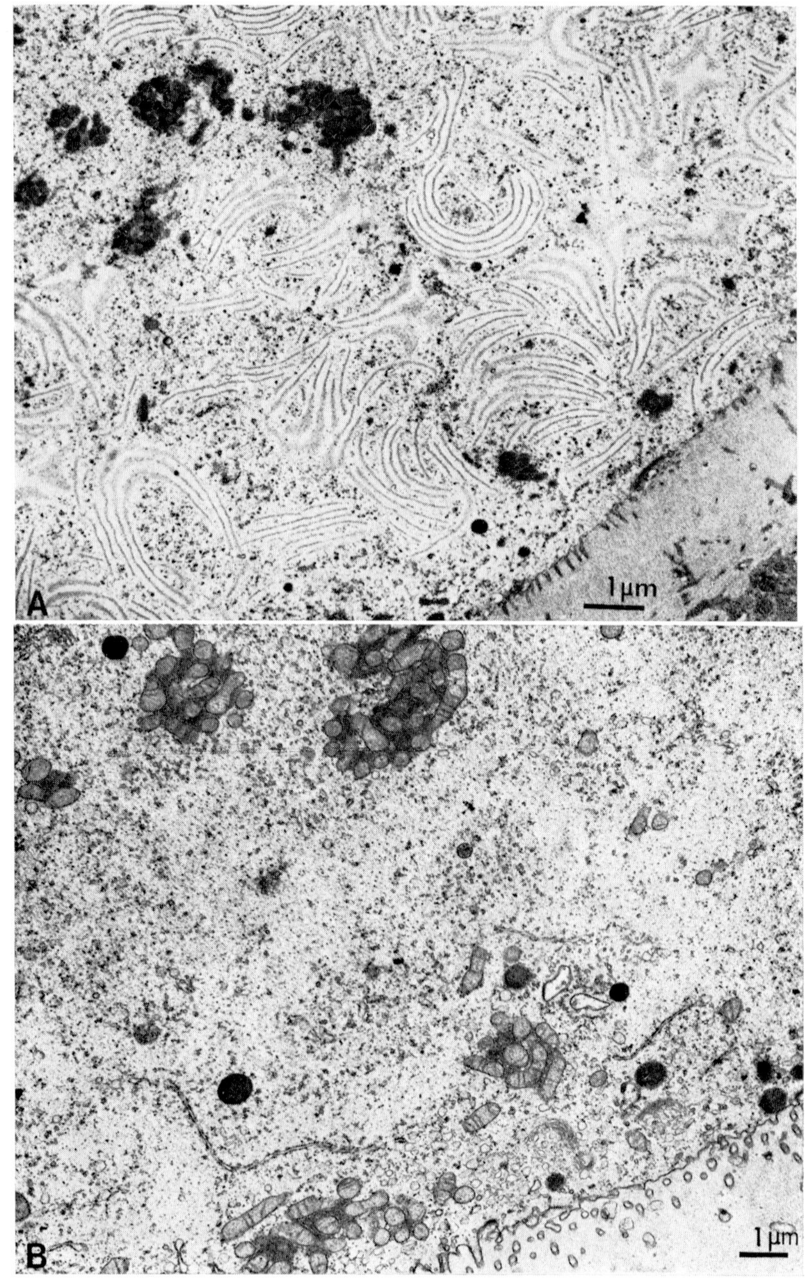

Plate 4. Material lost with one fixative but retained with another. (A) Peripheral cytoplasm of hamster oocyte fixed in glutaraldehyde followed by osmium tetroxide. The cytoplasm is filled with strands and whorls of material. Enzyme digestion experiments showed that this material consists of protein. (B) Peripheral cytoplasm of hamster oocyte fixed in osmium tetroxide. The cytoplasm between clusters of mitochondria is relatively empty except for ribosomes and small vesicles.

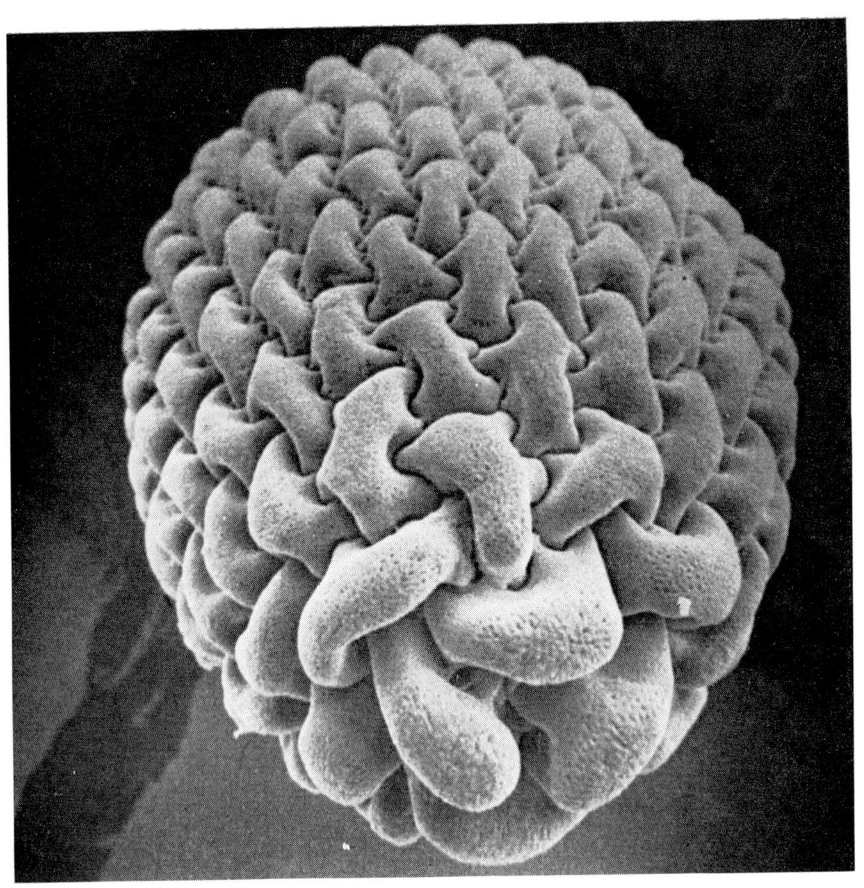

Plate 5. Scanning electron micrograph of egg of *Rhysotritia duplicata* (oribatid mite) ×1000. A Cambridge Stereoscan Micrograph reproduced by permission of The Pest Infestation Laboratories, Slough.

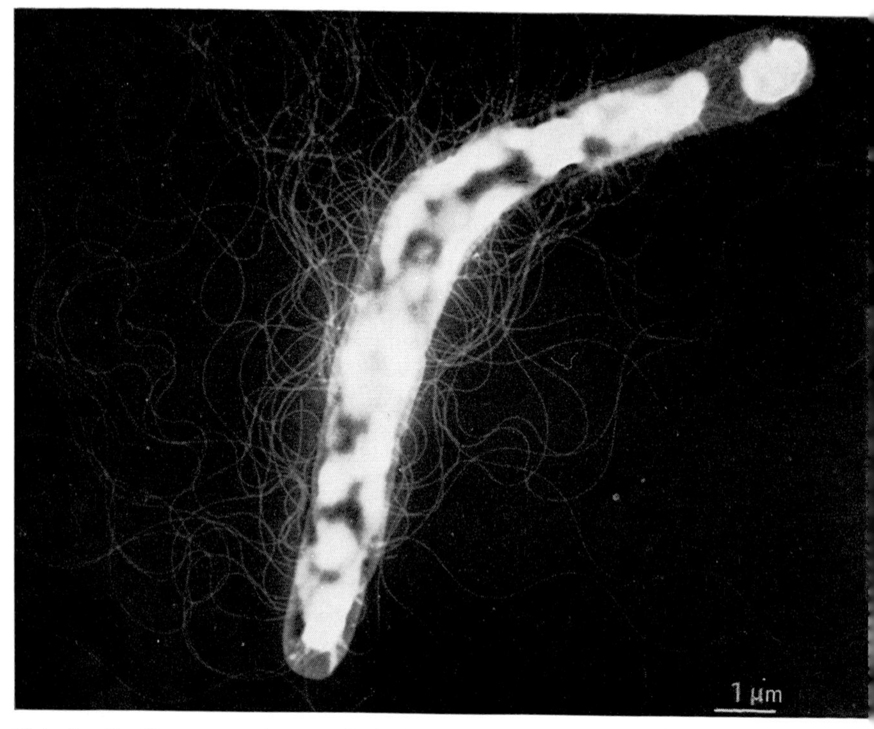

Plate 6. *Bacillus proteus* photographed with the million-volt electron microscope at Laboratoire d'Optique Electronique du C.N.R.S., Toulouse, France. Reproduced by permission of Dr G. Dupouy.

at the ultrastructural level may manifest themselves as any of the symptoms described under (*a*) to (*d*) below:

(*a*) *Swelling*. Swelling of cells and organelles can result if the tissue has been exposed to hypotonic solutions or if *post mortem* changes have occurred before fixation has been accomplished. In tissues where cell turnover is rapid, it may occur in a dying cell among healthy cells. It may also be symptomatic of certain disease processes, but if the animal from which the tissue was removed was healthy, preparative technique is the more likely cause. Mitochondria are particularly prone to this artefact. The average size of the mitochondria will be greater than the size characteristic for the tissue, and the mitochondrial matrix will be pale and perhaps patchy. The cristae may be enlarged (ballooned), or disorganized. Endoplasmic reticulum may also show swelling, particularly in dying cells. The cytoplasm may appear patchy or contain vacuoles.

It should be remembered, however, that mitochondria can vary in size with functional state, and that swollen endoplasmic reticulum in a secretory cell may simply represent storage of cellular products. Not all organelle enlargements can be summarily dismissed as artefacts.

(*b*) *Shrinkage*. This can occur if the tissue has been exposed to hypertonic solutions. One sign of this may be a scalloped nuclear outline. Mitochondria may be smaller than the average for the tissue, their outline uneven, and the matrix very dense. As with organelle enlargement, however, diminished size or difference in density are not always artificial.

(*c*) *Broken cell membranes*. If the cytoplasm of the cells is very pale and patchy, look carefully at the cell membranes. These may have ruptured and allowed leakage of cytoplasm into intercellular spaces. If this has occurred one not uncommonly sees cellular refuse in the intercellular spaces: bits of cytoplasm (not to be confused with the cytoplasmic processes of fibroblasts), a whorl of membranous material, an escaped mitochondrion. Rupture of the cell membranes can occur from osmotic insult or from mechanical damage incurred during handling of the tissue. If methacrylate embedding has been used it can also be caused by polymerization damage. It can occur, too, as the result of cell death.

(*d*) *Myelin figures*. At times whorls of membrane resembling the myelin which covers parts of the nervous system may be seen either within cells or lying in intercellular spaces. According to Trump and Ericsson (1965) and Curgy (1968) this is caused by partial breakdown of cellular lipoprotein membranes during aldehyde fixation. The lipid is then 'caught' by the osmium tetroxide used for post-fixation and fixed wherever it may happen to be. This explanation is probably valid in most cases. However, in certain developing tissues myelin figures may be seen at particular stages in the life history of a cell type and not at others. (Weakley, 1964, 1966; Coupland and Weakley, 1968). This holds true whether the tissue has been fixed in aldehyde followed by osmium tetroxide or in osmium tetroxide alone. This suggests that myelin figure formation may in some instances reflect developmental phenomena rather than fixation artefact.

(*e*) *Fixative-dependent differences in tissues.* It can never be assumed that the appearances of structures in a tissue will be the same if another fixative is used. One instance is the myelin figures mentioned above. Another is nuclear morphology. The nuclei of cells fixed with osmium tetroxide usually appear fairly homogeneous except for the nucleolus. Aldehyde fixation, however, generally results in the appearance of clumps of chromatin (DNA-containing nucleoprotein), particularly at the periphery of the nucleus. It is usually assumed that the dispersed state of the chromatin seen with osmium fixation more closely approximates the living state. This assumption is reinforced by the work of Ericsson and Biberfeld (1967) who demonstrated that when tissue is fixed by careful aldehyde perfusion chromatin clumping does not occur and that the nuclei resemble those fixed with osmium tetroxide.

(*f*) *Changes which mimic pathological alterations.* It should be noted that artefacts mentioned in (*a*) to (*e*) above can be caused not only by preparative techniques but by physiological processes, both normal and abnormal. This becomes very important in the study of pathological changes in tissues. If the tissue is not uniformly well fixed throughout the block, a whole range of cellular changes which mimic pathological changes will be encountered, as pointed out by Ericsson and Biberfeld (1967). Completely reliable criteria for assessing the many subtle variations which may occur have not as yet been established.

3. *Artefacts Introduced After Embedding*

(*a*) *Compression.* Even when ultrathin sections have been flattened with chloroform vapour a certain amount of compression caused by the cutting remains. This will make rounded objects appear flattened and should be taken into account when trying to make accurate measurements.

(*b*) *Dirt.* Dirt is quite easy to differentiate from tissue components, although its morphology is of course variable. It is the direct result of working in an unclean area or of sloppy technique.

It not only ruins otherwise good sections, but adds to the contamination of the microscope column.

(*c*) *Stain precipitate.* Stain precipitate is also easily differentiated from tissue components. Each stain has its characteristic form of aggregate. Uranyl acetate usually comes out as needle-like crystals of moderate electron density; lead citrate as rectangular crystals of extremely high electron density. (The shape may, of course, vary with staining conditions.) If you are having trouble with stain precipitate, refer to the staining instructions in the Appendix for advice.

(*d*) *Beam damage.* This has already been discussed in Chapter 5.

Measurement of Structures in Electron Micrographs

Since ångström units and micrometres are the two units of linear measurement most often used in electron microscopy it is convenient to remember that there is a difference of four decimal places between ångström units and micrometres: 2500 Å = 0·25 μm.

Measuring a Structure on a Negative or a Print

Suppose you photograph an unusual structure in the cytoplasm of a cell and you want to know how big it is. Before going further, remember that the profile which you see in your section is only a very thin slice through an object, the dimensions of which may vary tremendously if cut at other levels. Unless there are many of these structures present to be measured, your measurements may not tell you much about the real structure of the object.

Measurements may be made either on the negative or on a print taken from the negative. If a structure is very small it is more accurately measured if it is photographically enlarged. Also, reference lines can be drawn on the print, and the negative left intact for producing further prints if needed. If you measure from a print, remember that an error in determining the exact enlargement of the print will affect the accuracy of your final result. Also, the print may shrink slightly during processing.

An accurately calibrated ruler with half-millimetre divisions can be used for measuring larger objects. For measuring fine detail a special measuring magnifier incorporating a graticule should be used.

Deciding exactly where to measure may be difficult. If the structure appears 'round' its diameter will probably vary considerably if measured at a number of degrees of arc through an arbitrarily assigned 'central' point. A number of measurements will be necessary to determine an average diameter. If a 'round' body is slightly flattened, one must consider the possibility of compression during sectioning. Bodies of irregular shape, of course, present an even greater problem.

Approximate widths of elongate objects may be determined by averaging several measurements along their extent. Establishing length will be impossible if the object is longer than the thickness of the section and disappears out of it. The width of an object, e.g. a sheet of material, passing through a section can be accurately determined only if it passes through exactly at right angles. If it goes through obliquely, the apparent width will increase with the degree of obliquity and the edges will appear less dense than the centres (Fig. 13). The difficulty of determining with certainty

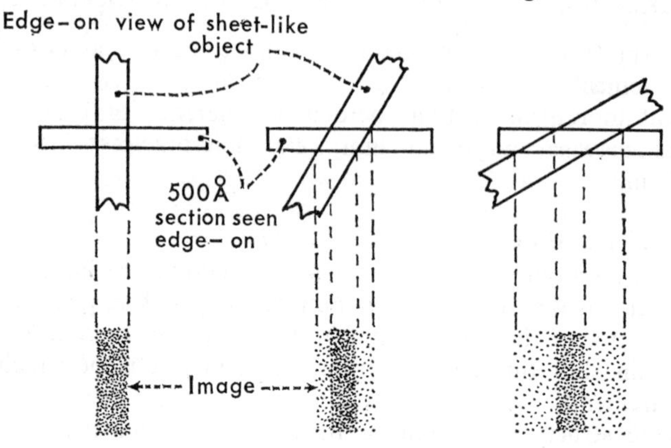

Fig. 13. Variation in apparent width of a sheet-like object passing at different angles through a section.

whether an object is passing at right angles or obliquely through a section will be more easily appreciated when you try to do it.

At the higher magnifications the problem becomes one of which two granular bumps to measure between. What appears to be black lines at low magnification become a collection of granules of varying densities at high magnification. A meaningful answer can only be obtained from many measurements carefully made and averaged.

After measurements have been made and averaged, calculations must be done to determine the actual size of the section through the structure which you have measured. This is simply a matter of dividing your average measurement by the product of the magnification at which the micrograph was taken and the photographic enlargement. In doing this, it is convenient to remember that there is a difference of seven decimal places from millimetres to ångström units. For example, if the object measures 10 mm and was photographed at a magnification of 40,000 and the micrograph enlarged five times: $40{,}000 \times 5 = 200{,}000$.

$$\frac{10\text{ mm}}{200{,}000} = \frac{100{,}000{,}000\text{ Å}}{200{,}000} = 500\text{ Å}$$

It must be remembered that the final figure incorporates a number of errors:

1. The magnification readings on the microscope are at best accurate to only ± 5 per cent. If measurements of extremely high precision are necessary the exact magnification produced by the electron microscope at each magnification setting must be calibrated, usually using a replica of a diffraction grating as a standard (See Meek, 1970 or Sjöstrand, 1967). Be sure that lens hysteresis has been taken into account as described on p. 82 before photographing.

2. Magnification will vary with the position of the specimen on the microscope axis. Make sure the grid is perfectly flat and has been inserted into the column with the specimen facing downward. Otherwise the position of the specimen can vary by the width of the grid.

3. At low magnifications unless the object was photographed at the centre of the field, distortion effects can alter the magnification.

4. Exact enlargement of the print may not have been accurately determined.

5. Some shrinkage of the print may occur while processing.

In short, conclusions drawn from measurements taken on an object photographed electron microscopically should be viewed with a certain scepticism, both in one's own work and when reading the literature.

Magnification Lines on Micrographs

The scale of a micrograph may be indicated by a labelled line in one corner. For instance, if the magnification is 50,000 the line could be made 50,000 μm (5 cm) long and labelled '1 μm', or 5 mm long and labelled 1000 Å.

Morphometry and Stereological Cytology

When all of the causes of error in measurement listed above have been taken into consideration, it should, above all, be remembered that what one is measuring is only a thin slice through a larger structure—so thin, in fact, as to be two-dimensional in practical terms. One cannot, of course, determine the three-dimensional structure of an organelle from a single 500 Å slice through it. A circular profile may be derived from any of the shapes in Figure 14.

The problem of three-dimensional structure must be attacked in statistical terms. If an organelle has certain features which enable one to recognize its profile it is possible, by taking a sufficiently large number of profiles into consideration, to draw certain conclusions about the three-dimensional characteristics and distribution of the organelle. To have statistical validity, samples must be both large and random, and collected from material which is truly representative of the tissue being studied.

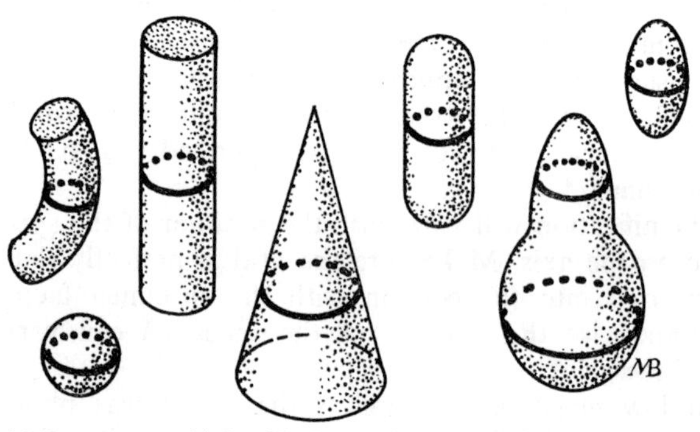

Fig. 14. Structures which will give a circular profile when sectioned.

The art of measuring organelles at the electron microscopic level so as to discover their three-dimensional structure and distribution in the cell has been developing rapidly in recent years. Weibel, Stäubli, Loud and others have shown that changes in the function of organelles, both normal and pathological, is often related to an increase or decrease in their numbers rather than to unambiguous morphological changes. It is hoped through morphometric and stereological methods to correlate biochemical and physiological findings with morphological ones.

The terms morphometry and stereology are defined by Weibel and Elias (1967) and Weibel (1969) as follows:

> '*Morphometry* implies the use of quantitative data in the description of structural features. Morphometric data can be obtained by a variety of measuring procedures performed on any type of specimen, but they can also be derived from stereological analysis of tissue sections. *Stereology* implies a geometric analysis of structures and textures; it includes methods that allow direct deri-

vation of metric properties of structures from 2-dimensional sections on the basis of geometrico-statistical reasoning.'

For a comprehensive review of the literature and a thorough discussion of the methods and techniques used in morphometry and stereology, see Weibel (1969).

Outlining a Research Project

It is all too easy for the beginner in any line of research to rush into a project with great enthusiasm, work diligently for several weeks and then realize that half of his techniques were either wrongly performed or misapplied, that his controls were inadequate, and that he kept no intelligible record of what he did.

If his research supervisor is as busy as many of them are, a student may well be left to get on with it as best he can. Perhaps the best protection against getting into a muddle is to start by making a complete outline of the research project. This should include the rationale and details of all techniques to be used, pertinent references to the literature and a list of supplies which will be needed. Also make a rough timetable. Use this as a guide only, not as a master. (Research usually takes longer than one expects by about a factor of five.) Get your supervisor to check your outline for feasibility before proceeding further.

A carefully made outline can be used later as an aid in writing a report, a paper for publication, or a thesis.

Keeping Records

The necessity of keeping records in any type of scientific endeavour is self-evident. Anyone seriously involved in research has far too many things on his mind to remember them all. And he doesn't need to prove his scientific adequacy by performing feats of memory.

Every experiment should be written down in detail. If you vary the recipe for a particular fixative, or the timing of a procedure, state how and why.

Great care should also be taken about marking tissues for purposes of identification. Nothing is more anonymous than a bit of blackened tissue in an araldite block—except perhaps a series of 500 Å sections on a copper grid dropped onto the floor along with half a dozen other grids. The most careful description in the world is meaningless if it has been done on the wrong specimen by mistake.

Keeping up With the Literature

Electron microscopy is a rapidly advancing field. New refinements in technique and equipment are constantly being introduced. The literature on ultrastructure is massive and increasing daily. If one is to do anything else besides read it is impossible to keep abreast of it. One should try to spend a couple of hours a week in a library scanning new issues of journals and following up references pertinent to one's work. A list of journals whose main concern is ultrastructure is given at the end of the chapter. *Current Contents,* a weekly publication which lists the titles of papers appearing in scientific journals, is a great help in locating recent publications of interest. It is to be found in most science libraries.

REFERENCES

General

Coupland, R. E. & Weakley, B. S. (1968). Developing chromaffin tissue in the rabbit: an electron microscopic study. *Journal of Anatomy* **102**, 425–455.

Curgy, J. J. (1968). Influence du mode de fixation sur la possibilité d'observer des structures myéliniques dans les hépatocytes d'embryons de poulet. *Journal de Microscopie* **7**, 63–80.

Fawcett, D. W. (1964). In *Modern Developments in Electron Microscopy*, Chap. 6 (Edited by B. M. Siegel), New York and London: Academic Press.

Harrison, G. A. (1971). Personal communication.

Meek, G. A. (1970). *Practical Electron Microscopy for Biologists*, Chap. 11 and Chap. 12, pp. 335–338, London and New York: Wiley-Interscience.

Sjöstrand, F. S. (1963). A new ultrastructural element of the membranes in mitochondria and some cytoplasmic membranes. *Journal of Ultrastructure Research* **9**, 340–361.

Sjöstrand, F. S. (1967). *Electron Microscopy of Cells and Tissues*, Vol. 1, Chap. 10, 11 and 12, New York and London: Academic Press.

Trump, B. F. & Ericsson, J. L. E. (1965). The effect of the fixative solution on the ultrastructure of cells and tissues. A comparative analysis with particular attention to the proximal convoluted tubule of the rat kidney. *Laboratory Investigation* **14**, 1245–1323.

Wachtel, A. W., Gettner, M. E. & Ornstein, L. (1966). In *Physical Techniques in Biological Research*, 2nd edn, Vol. 3, Part A, Chap. 4, New York and London: Academic Press.

Weakley, B. S. (1964). Ultrastructure of the fetal thymus in the golden hamster. *Journal of Morphology* **115**, 319–354.

Weakley, B. S. (1966). Electron microscopy of the oocyte and granulosa cells in the developing ovarian follicles of the golden hamster (*Mesocricetus auratus*). *Journal of Anatomy* **100**, 503–534.

Weibel, E. R. (1969). Techniques for quantitative evaluation of structure. 'Morphometry'. *International Review of Cytology* **26**, 235–302.

Weibel, E. R. & Elias, H. (1967). *Quantitative Methods in Morphology*, Berlin: Springer.

Atlases of fine structure

Fawcett, D. W. (1967). *An Atlas of Fine Structure—The Cell*, Philadelphia and London: W. B. Saunders.

Porter, K. R. & Bonneville, M. A. (1968). *Fine Structure of Cells and Tissues*, 3rd edn., London: Henry Kimpton.

Toner, P. G. & Carr, K. E. (1968). *Cell Structure*, Edinburgh and London: E. & S. Livingstone.

Interpretation of fine structure

Bahr, G. F. & Zeitler, E. H. (Eds.) (1965). Proceedings of a Symposium on Quantitative Electron Microscopy. *Laboratory Investigation* **14**, 739–1340.

Harris, R. J. C. (Ed.) (1962). *The Interpretation of Ultrastructure*, New York and London: Academic Press.

Haydon, G. B. (1968). On the interpretation of high resolution electron micrographs of macromolecules. *Journal of Ultrastructure Research* **25**, 349–361.

Haydon, G. B. (1969*a*). Electron phase and amplitude images of stained biological thin sections. *Journal of Microscopy* **89**, 73–82.

Haydon, G. B. (1969*b*). An electron-optical lens effect as a possible source of contrast in biological preparations. *Journal of Microscopy* **90**, 1–13.

Lenard, J. & Singer, S. J. (1968). Alterations of the conformation of proteins in red blood cell membranes and in solution by fixatives used in electron microscopy. *Journal of Cell Biology* **37**, 117–121.

McGee-Russell, S. M. & Ross, K. F. A. (Eds.) (1968). *Cell Structure and its Interpretation*, London: Edward Arnold.

Parsons, D. F. (Ed.) (1970). *Some Biological Techniques in Electron Microscopy*, New York and London: Academic Press.

Sjöstrand, F. S. (1969). In *Physical Techniques in Biological Research*, 2nd edn, Vol. 3C, pp. 169–200 (Edited by A. W. Pollister), New York and London: Academic Press.

Sjöstrand, F. S. & Barajas, L. (1968). Effect of modifications in conformation of protein molecules on structure of mitochondrial membranes. *Journal of Ultrastructure Research* **25**, 121–155.

Textbooks of cytology and molecular biology

Haggis, G. H., Michie, D., Muir, A. R., Roberts, K. B. & Walker, P. B. M. (1965). *Introduction to Molecular Biology*, London: Longmans.

Haggis, G. H. (1966). *The Electron Microscope in Molecular Biology*, London: Longmans, Green and Co.

Loewy, A. G. & Siekevitz, P. (1969). *Cell Structure and Function*. 2nd end., New York: Holt.

Robertis, E. D. P. de, Nowinski, W. W. & Saez, F. A. (1970). *Cell Biology*, 5th edn., Philadelphia and London: W. B. Saunders.

Textbooks in histology

Bloom, W. & Fawcett, D. W. (1968). *A Textbook of Histology*, 9th edn., Philadelphia and London: W. B. Saunders.

Ham, A. W. & Leeson, T. S. (1965). *Histology*, 5th edn., London: Pitman Medical Publishing. Philadelphia: J. B. Lippincott.

Journals specializing in ultrastructural and cytological studies

Journal of Ultrastructure Research. New York: Academic Press.

Journal of Cell Biology. New York: Rockefeller Press.
Journal de Microscopie. Paris: CNRS.
Zeitschrift für Zellforschung und microskopische Anatomie. Berlin and New York: Springer-Verlag.
Tissue and Cell. Edinburgh: Oliver and Boyd.
Journal of Microscopy. Oxford: Blackwells.
Experimental Cell Research. New York and London: Academic Press.
Journal of Cell Science. Cambridge: University Press.

Information on current publications

Garfield, E., (Ed.). *Current Contents: Life Sciences*, Philadelphia: Institute for Scientific Information, Inc. (Published weekly.)

8

Ultrastructural Cytochemistry

Electron cytochemistry is a rapidly expanding field which is still very close to its beginnings. Its development during the next decade will be multidirectional and extremely exciting for those involved. Already there are a number of relatively simple techniques which can be used by the beginner in electron microscopy, but he must be aware of the hazards of interpreting results obtained by methods which are still too new to have been fully evaluated.

I. INFORMATION TO BE GAINED FROM THE USE OF TWO OR MORE FIXATIVES

As has become evident from light microscopy, different fixatives affect cell components in different ways. Certain tentative information about the nature of a cell component may be gained from the way it responds to different fixatives. How different fixatives affect the subsequent staining of the component can also be a source of information. The tentative nature of this sort of information must, however, be stressed, and it should be used only as a preliminary to further investigations.

Potassium permanganate is little used now as a fixative in electron microscopy, since it extracts much of the protein and nucleoprotein from cells, leaving a skeleton composed largely of membranes, some carbohydrate and a little residual protein: not a particularly informative picture. Osmium tetroxide is a better protein fixative, but still much of the protein is extracted during fixation and the subsequent dehydration. Glutaraldehyde is the best protein fixative of all, because of its strong cross-linking ability.

The aldehydes stabilize protein, nucleoprotein and some carbohydrates, but allow lipids to be lost in the subsequent alcoholic dehydration. Thus if aldehyde fixation is used alone, lipid globules in cells will be seen as white areas ('negative images'). Cell membrane systems will also appear as negative images following aldehyde fixation if the usual uranyl acetate and lead stains are employed. However, some material remains at the site which may become visible after permanganate staining (Lewis and Knight, 1971). The extent to which various lipoprotein complexes ('masked lipids') are removed has not been accurately determined. The obvious example of such a complex is, of course, the membranes surrounding the cell and its organelles.

No obvious disintegration of the tissue is caused by the extraction of lipid, because the protein around or within the sites of lipid-containing material has been fixed and structural integrity of the cytoplasm is thus preserved.

If tissue fixed in glutaraldehyde is rinsed in buffer and then 'post-fixed' in osmium tetroxide, the osmium reacts with unsaturated lipids so that much less extraction occurs during dehydration. Thus the picture obtained at the end of processing is similar to that obtained by osmium fixation alone, except for the superior preservation of certain protein components.

Thus the simple use of two fixatives, alone and in combination, can give some tentative information of a cytochemical nature. If a structure appears as a negative image after aldehyde fixation alone, but is electron dense after post-fixation in osmium tetroxide, or after fixation with osmium tetroxide alone, it probably contains lipid.

If a structure is seen in tissue fixed in glutaraldehyde alone, and also in glutaraldehyde followed by osmium, but *not* in tissues fixed in osmium alone, it is likely to be either protein or carbohydrate. (See Plate 4A and B). The question of which it is can sometimes be resolved by the use of the enzyme digestion techniques outlined in a later section.

Nucleoproteins are preserved by both aldehyde and osmium fixation, so little information about them can be gained by the use of these two fixatives in conjunction. The state of aggregation of the nuclear chromatin is said to be different with aldehyde fixatives (clumped, denser) than it is with osmium tetroxide (pale, homogeneous), but exceptions are numerous. Table 5 summarizes the main information to be obtained by the use of the two most common fixatives. Plate 1 is pertinent.

II. LIMITATIONS OF POSITIVE STAINING

As pointed out in Chapter 4, specificity in a strict cytochemical sense is a term which can be applied to electron staining only with

Table 5. Appearance of various tissue components after certain fixatives (staining of sections with uranyl acetate followed by lead citrate is presumed)

	After glutaraldehyde alone	After osmium tetroxide alone	After glutaraldehyde followed by osmium
Lipid	Not present, lipid droplets and phospholipid membranes appear as negative images	Phospholipid membranes preserved, lipid droplets may appear complete or have washed-out centres	Appearance similar to that after osmium tetroxide alone
Protein	Well preserved, cytoplasm and mitochondrial matrix homogeneous and usually more electron-dense than after osmium tetroxide	Some protein lost, cytoplasm and mitochondrial matrix may be rather pale or patchy	Appearance similar to that after glutaraldehyde alone
Nucleic acids and nucleoprotein	Ribosomes preserved, chromatin may clump at nuclear periphery	Ribsosomes preserved. Chromatin less often clumped at nuclear periphery	Similar to appearance after glutaraldehyde alone
Glycogen	Preserved	Preserved best if Millonig's buffer is used	Said to be preserved best if Millonig's buffer is used in the post-fixing solution

extreme caution. The use of certain stains under rigidly specified conditions combined with adequate controls will sometimes yield quite precise information. For example, if a stain is known to react with more than one tissue component, blocking reagents may be used in the hope of blocking all the reactive sites save the one of interest. Another approach is the removal of certain tissue components which would otherwise react with the stain. This may be done with standard biochemical extraction procedures or by enzyme digestions as described in a later section. Enzyme digestion is preferable (if it can be made to work), since extraction procedures usually impair ultrastructural preservation.

III. LOCALIZATION OF SPECIFIC PROTEINS WITHIN THE TISSUES

Enzyme Localization

If an enzyme is present in active form within a tissue it can be made to act upon a substrate for which it is specific. The action of the enzyme will result in formation of a reaction product or products. If one of the reaction products can be made electron-dense it may be seen with the electron microscope.

Techniques for the localization of enzymes within tissues at the electron microscopic level have been largely adapted from the classical techniques of light histochemistry. They usually involve cutting 25 to 50 μm sections of fixed, frozen, unembedded tissue, since embedding media drastically hinder the activity of enzymes within the tissues. The sections are incubated in a solution which contains the specific substrate upon which the enzyme acts. The solution also contains some substance (often a heavy metal ion) which will react with one or more of the products of enzyme activity to form an insoluble electron-dense precipitate at the site of reaction. After incubation the sections may be post-osmicated if desired and then dehydrated and embedded in the routine manner.

As described in the section on freezing techniques in Chapter 10, a great deal of work has been done on methods of quick-freezing unfixed tissue in an endeavour to preserve tissue components unchanged and at the same time avoid ice-crystal damage. This approach has resulted in the accumulation of a considerable body of morphological information. However, the application of these methods to ultrastructural cytochemistry has not yielded the wealth of information originally hoped for. Even if the initial freezing procedure results in ultrastructural preservation of high quality, the technical difficulties of performing cytochemical reactions under conditions which can be expected to retain this preservation have been very difficult to overcome. Therefore, although fixation is always detrimental to enzyme activity, the tissue is in most cases fixed before freezing.

Fixation usually reduces enzyme activity in direct proportion to the cross-linking ability of the fixative. Unfortunately the quality of morphological preservation increases with the cross-linking ability of the fixative. Therefore a high degree of enzyme activity and good morphological preservation tend to be mutually exclusive. Considerable enzyme activity may remain after formaldehyde, which has a limited cross-linking ability. However, morphological preservation may be relatively poor and loss of protein high, especially since the tissue is more sensitive to freezing and sectioning damage after this fixative. Glutaraldehyde, which is a powerful cross-linking agent, gives better morphological preservation but enzyme activity may be low or absent. A reasonable compromise is frequently found in formaldehyde-fixed tissues, as long as the fixative is made up fresh from paraformaldehyde powder as recommended by Pease (1964). (See Appendix, p. 198.) Osmium tetroxide, although not a strongly cross-linking fixative, has generally been found unsuitable for enzyme work. Not only does it mask many reactive groups in the tissue, but osmium tends to denature enzymes.

Recently tissue sectioners for fixed unfrozen tissues have been developed, as described later. Use of these sectioners avoids freezing artefacts, and sectioning damage remains within reasonable limits.

A classical example of the localization of enzymes within the tissue is the Gomori technique for phosphatases. This has been adapted for use in electron microscopy by a number of workers (see references at the end of the chapter). Phosphatases are enzymes which split off phosphate groups. A phosphate-containing substrate (e.g. β-glycerophosphate) is incorporated into an incubation medium along with lead acetate. The tissue is incubated in this medium. The enzyme in the tissue attacks the substrate, splitting off its phosphate groups. The liberated phosphate ion then combines with the Pb^{2+} ion to form insoluble lead phosphate. The latter has a high electron density and is visible by electron microscopy, presumably at the site where the enzyme reaction has occurred.

The Gomori type of technique has resulted in the accumulation of a great deal of information concerning localization at the ultrastructural level of such enzymes as the acid and alkaline phosphatases and the nucleoside phosphatases and diphosphatases. Unfortunately, techniques of this sort are beset with a number of difficulties and the results are often questionable (Moses and Rosenthal, 1968; Ganote, Rosenthal, Moses and Tice, 1969; Etherton and Botham, 1970). Since fixation invariably cuts down enzyme activity and may abolish it altogether, a negative result means little or nothing. Penetration of the incubation medium into the tissue may present a major problem. Obviously the enzymes within cells cannot react with the substrate unless the substrate can reach the enzymes. If enzyme and substrate do meet and react, the products of the reaction are frequently diffusible. Thus they may either be lost from the tissue entirely or so displaced as to make determination of the original reaction site impossible. To add to the confusion, lead is particularly liable to adsorb on free surfaces and can give a completely spurious result as far as the *in vivo* location of enzyme activity is concerned. Careful controls are essential. However, it is generally conceded that certain positive reactions such as those for acid phosphates in lysosomes, ATPase in mitochondria, and nucleoside diphosphatase in the Golgi apparatus are valid, since biochemical techniques confirm the presence of these enzymes in the appropriate cell fractions.

Many attempts have been made to develop techniques for enzyme localization which avoid the pitfalls of Gomori-type reactions. Ogawa and Barrnett (1965) used tetrazolium salts to demonstrate the dehydrogenase system of enzymes in mitochondrial cristae. These salts act as electron acceptors in enzyme catalyzed oxidations, yielding electron-dense formazan particles. Although this method did not prove very useful, it is of interest since the reaction product is visible in the electron microscope even though the reagents used do not contain heavy metal atoms. Seligman *et al.* (1967) have improved the technique by using highly osmiophilic variants of tetrazolium salts.

Another method employing purely organic reagents is the simultaneous diazo-coupling technique adapted for electron microscopy by Lehrer and Ornstein, 1959). α-naphthyl acetate is used as the substrate for cholinesterase and the reaction product is coupled with hexazonium pararosanilin. The latter becomes visible after post-osmication. A somewhat more reliable method for cholinesterase was adapted for electron microscopy by Karnovsky (1964): Cholinesterase splits acetylthiocholine to form thiocholine. The thiocholine in turn reduces ferricyanide present in the incubation medium to ferrocyanide. The latter reacts with copper ions to produce electron-dense copper ferrocyanide.

A technique initiated by Karnovsky (1965) and Graham and Karnovsky (1966) and further developed by Novikoff and Goldfischer (1968, 1969), Fahimi (1969) and Seligman, Wasserkrug and Plapinger (1970), has been used extensively in the investigations of certain oxidative enzymes, including the catalases (peroxidases) and the cytochromes. The technique is based upon oxidation of diaminobenzidine (DAB) or related compounds by the enzymes. The reaction products which result are intensely osmiophilic.

So far, the enzymes which have proved amenable to study have been largely limited to those which catalyze hydrolytic or oxidative reactions. Investigation of the large number of enzymes which act in a synthetic rather than a catabolic capacity has scarcely begun.

For comprehensive reviews of enzyme localization techniques before 1965 and 1968 respectively, see Scarpelli (1965) and Bradbury (1968).

Immuno-Electron Microscopy

The antigen–antibody reaction is a highly specific affair. A particular antibody will react with one and only one particular antigen. A specific antibody may be produced by repeatedly injecting an animal with a particular antigen, and waiting until its system has produced antibodies programmed to react specifically with this antigen. The antibody-containing fraction of the animal's blood is then collected and the particular antibody of interest separated from the rest. It is then coupled with a substance which can be used as a 'marker' to trace the movements of the antibody in subsequent experiments.

In light microscopy the antibody is coupled to a fluorescent dye. The antibody–dye complex is then applied to living or fixed tissues and allowed to react with any specific antigen which may be present in the tissue. The tissue is then washed to remove antibody which has not reacted, and studied with a fluorescence microscope. Sites where an antigen–antibody reaction has occurred will fluoresce.

At the electron microscope level, the antibody may be labelled with an electron-dense marker which will cause the antigen–antibody reaction site to be visible in the electron microscope. Pioneer work in this field was done by Singer (1959) and Singer and Schick (1961) and carried further by several workers (for review see Sternberger, 1967). The Singer and Schick method involved the coupling of ferritin (an iron-containing electron-dense protein) to a specific antibody.

A major difficulty with the technique is that the ferritin–antibody complex seldom can be induced to penetrate into cells because of its large size. The reactions have thus far been largely confined to surface reactions. Another difficulty is that complete specificity of the preparation is hard to ensure. However, the technique is potentially capable of elucidating the early stages of antibody synthesis, determining the persistence of antigens, and the localization of specific proteins within cell organelles (Singer and Schick, 1961).

A more recent approach involves the attachment of an enzyme onto the antibody. This technique was independently proposed by Nakane and Pierce (1966) and Avrameas and Uriel (1966). For a review of this work see Avrameas (1970). After the enzyme-coupled antibody has reacted with the antigen in the tissue, a standard electron cytochemical technique for the localization of the enzyme can be performed. For electron microscopy the phosphatase and peroxidase enzymes are most commonly used, since fairly reliable methods have been devised for their localization. The penetration of the smaller antibody–enzyme complexes into the interior of cells is claimed to be somewhat better than with ferritin. It still presents a problem, however, since 7-S antibodies alone have an average molecular weight of 150,000 (Avrameas, 1970). In spite of its apparent potential, the technique has gained the reputation of being particularly difficult to perform satisfactorily.

IV. LOCALIZATION OF POLYSACCHARIDES AND MUCOSUBSTANCES

Polysaccharides in animal tissues serve as an energy reserve and also as components of the ground substance in connective tissue. Some of them combine with specific proteins to form mucopolysaccharides (which include the glycoproteins). Mucopolysaccharides are components of mucus secretions and of the cell coat (glycocalyx) which lies outside the plasma membrane of most cells. They are also present in basement membranes. Within cells they have been found to be associated with the Golgi apparatus, lysosomes, dense bodies, multivescular bodies and secretion or storage granules.

The presence of both polysaccharides and mucopolysaccharides can be detected at the electron microscopic level by using special techniques. Care must be taken, however, to ensure that loss of the material does not occur in processing the tissues.

The techniques which have been developed for detection of these compounds may be grouped into three classes:

1. *Methods Derived From the Classical Periodic Acid–Schiff Technique of Light Microscopy*

The periodic acid–Schiff technique will reveal polysaccharides and mucopolysaccharides by oxidizing their vic-glycol (1, 2-glycol) groups to aldehyde groups. The latter then react with Schiff's reagent.

For electron microscopy, Schiff's reagent is replaced by reagents which will cause an electron-dense deposit to form at the site of reaction. Three of these techniques are listed below with pertinent recent references. By consulting the extensive bibliographies of these papers the interested reader will be directed to earlier work relative to the techniques.

A. The periodic acid–silver methenamine technique (Rambourg, 1967*a*; Rambourg, Hernandez and Leblond, 1969).

B. The periodic acid–thiocarbohydrazide–osmium tetroxide technique (Seligman, Hanker, Wasserkrug, Dmochowski and Katzoff, 1965; Thiery, 1967).

C. The periodic acid–thiocarbohydrazide (or thiosemicarbazide)–silver proteinate technique (Thiery, 1967; Flechon, 1970).

D. The pentafluorophenylhydrazine technique (Bradbury and Stoward, 1967).

These techniques presumably demonstrate the same range of carbohydrate substances as does the classical PAS reaction. Controls omit either the periodic acid or the 'staining' reagent. The thiocarbohydrazide and pentafluorophenyl hydrazine techniques appear to be the most promising. A substance containing a thiol (SH) group is linked to the carbohydrate molecule. The thiol group can then react with silver or osmium.

2. *Methods Based on the Affinity of Phosphotungstic Acid at Low pH for Polysaccharides*

The mechanism of staining and the specificity of the PTA reaction for polysaccharide is still highly controversial, as a glance at two juxtaposed articles in a recent issue of *Journal of Histochemistry and Cytochemistry* will attest (Glick and Scott, 1970; Pease, 1970). See also Scott and Glick (1971).

3. *Methods Specific for Acid Groups*

These methods are based on classical techniques for demonstration of polyanions rich in sulphate or carboxyl groups. They are used at the electron microscope level to demonstrate acid mucopolysaccharides.

A. *The colloidal thorium dioxide technique of Revel* (*1964*). The thorium dioxide presumably binds to the uronic, carboxyl or sulphate groups on acid mucopolysaccharides. The reaction is performed on ultrathin methacrylate sections. A modification by Rambourg and Leblond (1967) is performed on small blocks of fixed tissue.

B. *Methods based on Hale's colloidal iron technique* (Wetzel, Wetzel and Spicer, 1966; Gasic, Berwick and Sorrentio, 1968). The reaction is performed on 25 to 50 μm sections cut on a freezing microtome, and penetration into the tissue tends to be poor. The positively charged colloidal iron particles presumably attach to uronic, carboxyl or sulphate groups on the acid mucopolysaccharides.

C. *Ruthenium red staining.* This technique was introduced by

Luft (1966, 1968). He added the stain to the fixative, but found penetration to be so poor that only the mucopolysaccharides on cell surfaces or in intercellular spaces were stained. In order to facilitate penetration, Gustafson and Pihl (1967) adapted the technique for use on ultrathin sections of glycol methacrylate-embedded tissue. They obtained intense staining of mast cell granules, and some nuclear staining under certain conditions. The usefulness of this stain for ultrathin sections has yet to be fully investigated.

D. *Alcian blue*. It has been reported by Behnke and Zelander (1970) that one per cent alcian blue added to glutaraldehyde fixative will render acid mucosubstances and glycoproteins insoluble. The resulting dye–mucosubstance complex is osmiophilic, and therefore visible in the electron microscope after the tissue has been post-fixed with osmium tetroxide. Since the stain is added to the fixative rather than being used on ultrathin sections, penetration problems have confined the usefulness of this stain largely to cell exteriors and intercellular substance.

Staining of ultrathin sections with alcian blue was reported by Tice and Barrnett (1962) and more recently by Rothman (1969) but specificity for mucosubstances has not been claimed under these conditions.

E. *Lanthanum*. The action of lanthanum is essentially similar to ruthenium red and alcian blue in being confined to cell surfaces and intercellular material. It has been used to test the permeability and reveal the structure of epithelial 'tight junctions' (Revel and Karnovsky, 1967) and has been shown to precipitate protein–polysaccharide extracted from cartilage (Khan and Overton, 1970). Its ability to bind to specific extracellular components appears to depend upon the particular conditions under which it is used. When employed during permanganate fixation it binds strongly to a 50 Å wide surface layer which is continuous with the plasma membrane and which is removed by phospholipase (Lesseps, 1967). For a review of lanthanum staining see Martinez-Palomo (1970).

VI. ENZYME DIGESTION TECHNIQUES

These methods are useful principally in a confirmatory capacity, and are performed in conjunction with electron staining or other cytochemical techniques.

If an enzyme which specifically controls the breakdown of a particular substance is introduced into a tissue containing that substance, the substance will be broken down ('digested') and may then diffuse out of the tissue and be lost. By comparing sections of tissue which have been exposed to the enzyme with sections of tissue which have not been exposed, one can sometimes determine the morphological locations where material has been removed. If these locations routinely correspond to parts of a certain organelle, it is fairly safe to suggest that the organelle normally contains the substance.

Scrupulous adherence to proper techniques for handling enzymes is essential to success. The enzyme must be of proven activity (by some pertinent test *in vitro*) and of sufficiently high purity to be specific—i.e. it must be specified by the manufacturer to be uncontaminated by other enzymes. It must be stored with a drying agent in a deep freeze or refrigerator freezing compartment when not being used. Glassware must be specially cleaned (see Appendix, p. 216). Water used for making up the incubation media must be distilled twice in a glass still (i.e. a still containing no metal parts) in order to avoid inactivation of the enzyme by metal ions. The enzyme should be weighed using glass implements rather than metal ones.

There are two principal methods of performing enzyme digestions on tissues to be viewed with the electron microscope. In both of them, choice of fixative is important. All fixatives reduce enzyme activity to some extent. Some enzymes will act only on tissues fixed with formaldehyde; others will act after fixation with any of several aldehydes; very few will act after osmium tetroxide. Enzyme activity is usually greatest after formaldehyde, but tissue preservation is generally better after glutaraldehyde. Therefore if the enzyme is active after both fixatives, glutaraldehyde is usually the fixative of choice. The embedding material may also affect the activity of the enzyme when digestions are done on sections of embedded material.

1. *Digestion of Small Blocks or Thick Sections of Unembedded Tissue*

This method involves incubation of small blocks ($\frac{1}{2}$ to 1 mm on edge) or sections of fixed tissues (usually 25 to 50 μm in thickness) in a medium containing the digesting enzyme. After incubation, the blocks or sections are rinsed, dehydrated and embedded. If aldehyde fixation is used, the tissues may be post-fixed in osmium tetroxide after incubation has been completed.

Small blocks are easily prepared and handled, but penetration of the enzyme into the block is usually poor and digestion may be limited to the first layer or two of cells. Better penetration is usually obtained by using frozen sections 25 to 50 μm thick cut from small blocks of fixed tissue on a cryostat or a CO_2 freezing microtome. The sections are then collected in ice-cold buffer of the same composition as that used in the fixative, or in a sucrose solution. These sections are subsequently incubated in medium containing the enzyme.

With the frozen section method, however, there may be considerable damage to the tissue because of the freezing and thawing involved and the passing of the knife through the frozen tissue during sectioning. It is very important that the knife be freshly sharpened and in perfect condition.

A method for obtaining 20 to 50 μm sections without freezing the tissue was introduced by Smith and Farquhar (1965). Their tissue chopper allows uniform sections of fixed material to be cut in the range of 20 μm to 100 μm, and morphological preservation and enzyme localization on such sections is reported to be superior to that with frozen sections. More recently, another type of sectioner has been produced which permits cutting of unfrozen sections in the range of 5 to 20 μm. A comparison and evaluation of the two types of sectioner was made by Smith (1970).

Whether small blocks or sections are used, the subsequent incubation, post-fixation, dehydration and embedding procedures are the same. The sections should, of course, be flat-embedded in embedding boats rather than in capsules.

Once the 20 to 50 μm sections have been embedded, obtaining ultrathin sections from them may present some difficulty. Incubated tissue is often difficult to cut, and sectioning an entire 20 μm slice without obtaining a usable ultrathin section is not uncommon. It is wise, therefore, to incubate and embed a fairly large number of sections.

The time of incubation varies with the enzyme, the fixative, the temperature and the tissue. There are so many variables that trial and error is usually necessary, although some idea may be obtained from reading the pertinent literature. One can try removing tissue from the incubating medium after 5 min, 15 min, 30 min, 1 hour and 2 hours. The tissue must be carefully rinsed in buffer several times to ensure that the enzyme has been removed. Control sections are treated in exactly the same manner except that the incubation medium either contains no enzyme at all, or has been treated with an enzyme inhibitor. Careful controls are especially necessary in enzyme digestion techniques, because of the possibility that the material in question might be lost into the surrounding medium during extended incubation even if no enzyme were present. *Great care must be used not to contaminate the control solutions with enzyme.* This could happen by inadvertently using the same brush for transferring both experimental and control sections from one solution to another. Even minute traces of enzyme may render the control material uninterpretable.

If the technique being used calls for incubation at 37°C it is worthwhile incubating some tissue at room temperature as well, since preservation of the tissue ultrastructure will be better at the lower temperature and the enzyme may still be sufficiently active to give the desired results.

One-micrometre sections taken from the 25 to 50 μm embedded sections may also be informative about the extent of enzyme digestion of the tissues, and should be carefully studied.

2. *Digestion of Ultrathin Sections of Tissue*

Even when one uses 25 to 50 μm sections, penetration of the enzyme into the tissue may still be a problem. A negative result obtained using such material may not be significant. For this reason attempts have been made for years to carry out digestions on ultrathin sections of embedded tissue. Unfortunately, the presence of embedding medium tends to hinder penetration of the enzyme despite the extreme thinness of the sections.

The tissue is processed in the usual manner when epoxy embedding is used. If water-soluble embedding media are used, dehydration is usually accomplished in increasing percentages of the embedding medium itself, rather than in ethanol. It may also be accomplished by 'inert dehydration' of tissues preserved and dehydrated in glycol as described by Pease (1966*a* and *b*). With the methacrylates, both water-soluble and non-water-soluble, polymerization of the embedding material has been found to proceed with less damage to the tissue if it is done in a cold room under an ultraviolet lamp (Leduc and Bernhard, 1967). Osmium tetroxide cannot be used as a fixative in this case, since it renders the tissue opaque to ultraviolet light and polymerization will not proceed to completion within the tissue.

The ultrathin sections are incubated in a medium containing the desired enzyme. Incubating sections mounted on copper grids is to be avoided because the metal ions from the grid may inactivate the enzyme.

Passing unmounted ultrathin sections through a series of solutions and getting them safely onto a grid at the end of the exercise can be a hair-raising if not a defeating experience. For practical methods of doing this see Marinozzi (1964) and McGee Russell (1965). An alternative is to use gold grids, as gold is far less reactive than copper and can in most cases be used without interfering with the enzyme. Gold grids are very expensive; they are also very flexible and difficult to handle. Most laboratories encourage their personnel to perfect the technique of processing unmounted sections.

For all methacrylate sections (both water-soluble and non-water-soluble) the grids should be coated with a thin film of collodion or Formvar reinforced by a thin layer of carbon (see Appendix, p. 215), as unsupported sections tend to break up when exposed to the electron beam.

Results of enzyme digestions on ultrathin sections are extremely variable and findings obtained in one laboratory are often unrepeatable in another. Success has been more often achieved with the water-soluble embedding media, although morphological preservation is not usually of as high quality as with material embedded in epoxy resins. Also, obtaining good sections is often difficult with water-soluble media.

Water-soluble embedding media were introduced in the early 1960's: glycol methacrylate (GMA) by Rosenberg, Bartl and Lěsko (1960); Durcupan by Stäubli (1960); hydroxypropyl methacrylate (HPMA) by Leduc and Holt (1965). GMA and HPMA have been the most useful for enzyme work, and the modification of the GMA embedding technique by Leduc and Bernhard (1967) is usually the method of choice. The use of these media is almost exclusively limited to enzyme digestions, enzyme localizations being generally unsuccessful (Bernhard, 1966).

Both proteases and nucleases are active on formalin or formol-acrolein fixed tissues embedded in glycol methacrylate (Leduc and Bernhard, 1962; Leduc, Marinozzi and Bernhard, 1963; Marinozzi and Bernhard, 1965). Proteases such as pepsin and pronase are also active on glutaraldehyde-fixed material. The activity of the nucleases is more variable, and some workers report that nucleoprotein removal can only be achieved if protease digestion is carried out first. Ribonuclease is active after both formaldehyde and glutaraldehyde. Deoxyribonuclease is usually unsuccessful after glutaraldehyde fixation, and in any case, gives inconsistent and questionable results whatever fixative is used. It shows no activity whatever after HPMA embedding.

Until the work of Monneron and Bernhard (1966) it was generally believed that material embedded in epoxy resins could not be attacked by enzymes. These authors showed that some removal of protein and polysaccharide is possible on sections embedded in Epon. The electron density of ribosomes is diminished but not removed; nucleolar RNA appears unaltered. Attempts to remove DNA were unsuccessful. However, Douglas (1970) has reported removal of DNA from ultrathin sections of Epon–Araldite-embedded tissue with perchloric acid, so this appears at present to be the method of choice. A mixture of Epon and Araldite has also been used with some of the proteases (Anderson and Ellis, 1965; Anderson and André, 1968; Douglas, Ripley and Ellis, 1970).

The difficulties encountered at the electron microscope level in removing nucleoprotein are thought to be due to the binding of the nuclease to the nucleoprotein (Aldridge, 1962; Alfert, Dass and Eastwood, 1962; Moreno and Venyins, 1970). This binding results in the abolition of specific staining at the light microscopic level because the reactive sites are blocked by the nuclease. Thus the nucleoprotein appears to have been 'removed' since it no longer stains. At the electron microscope level, however, the binding of the nuclease to the nucleic acid can actually result in an increase in electron density unless the ribonucleoprotein is first split apart by the use of a protease.

References to key papers in the field of enzyme extraction are given at the end of the chapter. Further references will be found in the bibliographies of these papers.

VII. CORRELATIVE CYTOCHEMICAL WORK WITH 1 μm SECTIONS FOR LIGHT MICROSCOPY

Staining of 1 to 2 μm sections of methacrylate-embedded tissues for light microscopy became common practice in the early days of sectioning for electron microscopy. It was not at first appreciated that the tissue could be stained without the embedding material being removed first. Methacrylate can be removed from sections with xylene, but this tends to be detrimental to the tissues.

When epoxy resins were first introduced it was generally assumed that staining with the ordinary stains used in light microscopy would be impractical since the epoxy embedding material could not be removed except by drastic methods. Lane and Europa (1965) found that a saturated solution of sodium hydroxide in absolute ethanol would remove the resin, but staining reactions in the remaining tissue were, not surprisingly, rather disappointing.

It soon became apparent, however, that staining could be accomplished without removal of the epoxy resin as long as certain conditions were used. Useful staining procedures were developed in a number of laboratories and several of these are referred to by Glauert (1965). Stevens (1966) gives an extensive list of publications dealing with stains for tissues embedded in methacrylate, Vestopal W, Araldite and Epon, but there appears to have been no comprehensive review of developments since that time.

Appendix VII gives recipes for several stains which have been found useful. It must be remembered, however, that any specificity or selectivity which these stains show when applied to tissue fixed with ordinary histological fixatives and embedded in paraffin may be lost entirely or altered in unexpected ways when applied to epoxy-embedded material. For the moment, the great virtue of staining 1 to 2 μm epoxy sections (aside from the obvious primary value of locating tissue for thin sectioning) is that alternating 1 μm and ultrathin sections can be compared to give a wealth of additional information about the tissue being studied. Many cytochemical reactions which are observed with the electron microscope are also visible in 1 μm sections, and cellular detail is far superior to that seen with 5 μm sections of ordinary histological material.

VIII. NEW DIRECTIONS IN ULTRASTRUCTURAL CYTOCHEMISTRY

The techniques described in this section will certainly not be performed by the beginner in electron microscopy. They are included simply to indicate where new ground is being broken.

Cryo-Ultramicrotomy

For the past few years attempts have been made to perfect a technique for cutting ultrathin sections of unembedded frozen material. The impetus for doing this has come mainly from histochemists and cytochemists endeavouring to determine the intracellular location of particular enzymes. As has been made clear in the present chapter, several problems hinder cytochemical work with the present methods used in electron microscopy:

1. Loss of material from the cells and tissues during processing.
2. Displacement of material from its original position within the tissue.
3. Chemical alteration of macromolecular structure during fixation and subsequent processing, with consequent loss of biological activity.
4. Difficulties in penetration of cytochemical reagents into thick sections of unembedded material or into ultrathin sections of embedded material.

If it were possible to freeze unfixed tissue instantaneously so that all metabolic processes ceased and no movement or loss of material could occur, and then to obtain *ultrathin* frozen sections upon which to perform cytochemical reactions, these problems would be solved. In practice this ideal has not yet been achieved, although intensive work is being done in several laboratories which is bringing its realization closer.

Preliminary attempts at ultrathin frozen sectioning (Fernandez-Moran, 1952) were disappointing, and were suspended in favour of the more promising sectioning of methacrylate-embedded material. Bernhard's group in France rekindled interest in this area, and a technique was developed (Bernard *et al*, 1967; Leduc *et al*, 1967) which has enabled various cytochemical tests to be performed on ultrathin sections of fixed frozen tissues.

Promising results with the sectioning of *unfixed*, unembedded frozen tissues have been coming from a number of laboratories (Christensen, 1967, 1968; Appleton, 1968). Hodson and Marshall (1969) have developed an 'ultracryotome' which will cut ultrathin frozen sections of unfixed, unembedded specimens maintained at the temperature of liquid nitrogen. The tissue is frozen in dichlorodifluoro-methane cooled to its freezing point by liquid nitrogen. Sections are cut with a glass knife onto cyclohexane and collected on carbon-coated copper grids. The difficulty of collecting the sections on the grid was neatly solved: An electrically charged copper grid is partially immersed in the cyclohexane bath where the sections are floating. Excess charge streams from the grid and charges the floating sections. Then the potential at the grid is lowered relative to the frozen sections. The sections are attracted to the grid and spread themselves over it.

The grids are then transferred into the freeze-drying chamber of the ultramicrotome where the ice is removed from the sections by sublimation. The chamber is then allowed to reach room temperature and the sections are ready to be viewed.

This technique is potentially capable of solving all four of the cytochemical problems listed above. The only chemical with which the tissue comes into contact (other than the coolant) is cyclohexane; at the temperatures employed it does not appear to remove lipid or other materials from the tissues.

The Analytical Electron Microscope

One of the most exciting recent developments in the field of electron microscopy is the combination of the electron probe X-ray microanalyzer and the high resolution transmission electron microscope.

The electron-probe microanalyzer scans the specimen with a thin pencil of electrons (the probe). Interaction of the probe with the specimen produces X-ray emission which can be analyzed spectrometrically to identify chemical elements present in the specimen.

The EMMA-4 analytical electron microscope recently made

commercially available by AEI combines in one machine an efficient electron probe microanalyzer with an electron microscope capable of 10 Å resolution. An ultrathin section may be studied in the usual high resolution manner. When an area of the specimen is found which requires chemical analysis, the analysis can immediately be performed without moving the specimen or losing the particular field of view. One simply turns on the probe lens and focuses the probe. Analysis is possible of areas measuring from slightly less than one thousand ångströms in diameter to a few thousand ångströms. Determination of the amount of material present can be made from the number of X-ray quanta per second counted while the material is under the probe.

If no prior knowledge is available about the chemical elements contained in the specimen, the instrument may be set to scan automatically through the periodic table and trace out a spectrum on a chart recorder. Alternatively, particular elements may be located accurately in the specimen by tuning the spectrometer to a particular element and making point by point analyses across the section. The probe may, at any time, be defocused and the area of interest studied in the normal high resolution manner.

Obviously this instrument has immense potential in a number of fields. Notable among these are histochemistry and cytochemistry. Here, analytical electron microscopy may soon render obsolete the present unsatisfactory system of using electron stains of doubtful specificity or enzyme techniques which are unrepeatable from one laboratory to another. Specific staining methods can now be used, the reactions of which were formerly detectable only at the level of the light microscope. With appropriate labelling of substances for detection by the probe, transfer of metabolites within and between cells may be traced. Pioneer work with EMMA-4 is already being done using ultrathin frozen sections of unfixed tissue described in the preceding section.

The medical applications are open-ended. With the analytical electron microscope it will be possible to determine the natural mineral content of tissues, detect the presence of foreign material, and trace the movements of the latter (Robertson, 1968). Once normal baselines for the various tissues and cells have been established with the analytical electron microscope, previously obscure causes for pathological tissue changes may be identified, or disease detected at stages where as yet no symptoms are present in the patient.

REFERENCES

Fixation

Lewis, P. R. & Knight, D. (1971). Personal communication.

Enzyme localization techniques

1. General

Bradbury, S. (1968). In *Cell Structure and its Interpretation*, pp. 135–147 (Edited by S. M. McGee-Russell and K. F. A. Ross), London: Edward Arnold.

Pease, D. C. (1964). *Histological Techniques for Electron Microscopy*, 2nd edn., p. 53, New York and London: Academic Press.

Scarpelli, D. G. & Kanczak, N. M. (1965). Ultrastructural cytochemistry: principles, limitations, and applications. *International Review of Experimental Pathology* **4**, 55–126.

2. Gomori type reactions

Etherton, J. E. & Botham, C. M. (1970). Factors affecting lead capture methods for the fine localization of rat lung acid phosphatase. *Histochemical Journal* **2**, 507–519.

Ganote, C. E., Rosenthal, A. S., Moses, H. L. & Tice, L. W. (1969). Lead and phosphate as sources of artifact in nucleoside phosphatase histochemistry. *Journal of Histochemistry and Cytochemistry* **17**, 641–650.

Goldfischer, S., Essner, E. & Novikoff, A. B. (1964). The localization of phosphatase activities at the level of ultrastructure. *Journal of Histochemistry and Cytochemistry* **12**, 72–95.

Holt, S. J. & Hicks, R. M. (1962). In *Interpretation of Ultrastructure*, pp. 193–211 (Edited by R. J. C. Harris). New York and London: Academic Press.

Moses, H. L. & Rosenthal, A. S. (1968). Pitfalls in the use of lead ion for histochemical localization of nucleoside phosphatase. *Journal of Histochemistry and Cytochemistry* **16**, 530–539.

Novikoff, A. B., Essner, E., Goldfischer, S. & Heus, M. (1962). In *The Interpretation of Ultrastructure*, pp. 149–192 (Edited by R. J. C. Harris), New York and London: Academic Press.

Sabatini, D. D., Bensch, K. & Barrnett, R. J. (1963). Cytochemistry and electron microscopy. The preservation of cellular ultrastructure and enzymatic activity by aldehyde fixation. *Journal of Cell Biology* **17**, 19–58.

Sabatini, D. D., Miller, F. & Barrnett, R. J. (1964). Aldehyde fixation for morphological and enzyme histochemical studies with the electron microscope. *Journal of Histochemistry and Cytochemistry* **12**, 57–71.

Wachstein, M. & Meisel, E. (1957). Histochemistry of hepatic phosphatases at a physiologic pH with special reference to the demonstration of bile canaliculi. *American Journal of Clinical Pathology* **27**, 13–23.

3. Techniques for the dehydrogenases

Ogawa, K. & Barrnett, R. J. (1965). Electron cytochemical studies of succinic dehydrogenase and dihydronicotinamide–adenine dinucleotide diaphorase studies. *Journal of Ultrastructure Research* **12**, 488–508.

Seligman, A. M., Ueno, H., Morizono, Y., Wasserkrug, H. L., Katzoff, L.

& Hanker, J. S. (1967). Electron microscopic demonstration of dehydrogenase activity with a new osmiophilic ditetrazolium salt (TC-NBT). *Journal of Histochemistry and Cytochemistry* **15**, 1–13.

4. Diazo coupling and cholinesterase

Karnovsky, M. J. (1964). The localization of cholinesterase activity in rat cardiac muscle by electron microscopy. *Journal of Cell Biology* **23**, 217–232.

Lehrer, G. M. & Ornstein, L. (1959). A diazo coupling method for the electron microscopic localization of cholinesterase. *Journal of Biophysical and Biochemical Cytology* **6**, 399–404.

Tennyson, V. M. & Brzin, M. (1970). The appearance of acetylcholinesterase in the dorsal root neuroblast of the rabbit embryo. *Journal of Cell Biology* **46**, 64–80.

5. Reactions for peroxidase (including catalase) and the cytochromes

Bainton, D. F. & Farquhar, M. G. (1970). Segregation and packaging of granule enzymes in eosinophilic leukocytes. *Journal of Cell Biology* **45**, 54–48.

Fahimi, H. D. (1969). Cytochemical localization of peroxidatic activity of catalase in rat hepatic microbodies (peroxisomes). *Journal of Cell Biology* **43**, 275–288.

Graham, R. C. & Karnovsky, M. J. (1966). The early stages of absorption of injected horseradish peroxidase in the proximal tubule of the mouse kidney. Ultrastructural cytochemistry by a new technique. *Journal of Histochemistry and Cytochemistry* **14**, 291–302.

Karnovsky, M. J. (1965). Vesicular transport of exogenous peroxidase across capillary endothelium into the T system of muscle. *Journal of Cell Biology* **27**, 49A–50A.

Legg, G. & Wood, R. L. (1970). New observations on microbodies. A cytochemical study on CPIB-treated rat liver. *Journal of Cell Biology* **45**, 119–129.

Novikoff, A. B. & Goldfischer, S. (1968). Visualization of microbodies for light and electron microscopy. *Journal of Histochemistry and Cytochemistry* **16**, 507.

Novikoff, A. B. and Goldfischer, S. (1969). Visualization of peroxisomes (microbodies) and mitochondria with diaminobenzidine. *Journal of Histochemistry and Cytochemistry* **17**, 675–680.

Seligman, A. M., Wasserkrug, H. L. & Plapinger, R. E. (1970). Comparison of ultrastructural demonstration of cytochrome oxidase activity with three bis (phenylene-diamines). *Histochemie* **23**, 63–70.

Immuno-electron microscopy

Avrameas, S. (1970). Immunoenzyme techniques: Enzymes as markers for the localization of antigens and antibodies. *International Review of Cytology* **27**, 349–385.

Nakane, P. K. & Pierce, G. B. (1966). Enzyme-labeled antibodies: preparation and application for the localization of antigens. *Journal of Histochemistry and Cytochemistry* **14**, 929–931.

Singer, S. J. (1959). Preparation of an electron-dense antibody conjugate. *Nature, London* **183**, 1523–1524.

Singer, S. J. & Schick, A. F. (1961). The properties of specific stains for electron microscopy prepared by the conjugation of antibody molecules with ferritin. *Journal of Biophysical and Biochemical Cytology* **9**, 519–537.

Sternberger, L. A. (1967). Electron microscopic immunocytochemistry; a review. *Journal of Histochemistry and Cytochemistry* **15**, 139–159.

Techniques for polysaccharides and mucosubstances

Behnke, O. & Zelander, T. (1970). Preservation of intercellular substance by the cationic dye alcian blue in preparative procedures for electron microscopy. *Journal of Ultrastructure Research* **31**, 424–438.

Bradbury, S. & Stoward, P. J. (1967). The specific cytochemical demonstration in the electron microscope of periodate-reactive mucosubstances and polysaccharides containing vic-glycol groups. *Histochemie* **11**, 71–80.

Flechon, J.-E. (1970). Nature glycoprotéique des granules corticaux de l'oeuf de lapine. Mise en evidence par l'utilisation comparée de techniques cytochimiques ultrastructurales. *Journal de Microscopie* **9**, 221–242.

Gasic, G. J., Berwick, L. & Sorrentio, M. (1968). Positive and negative collodial iron as cell surface electron stains. *Laboratory Investigation* **18**, 63–71.

Glick, D. & Scott, J. E. (1970). Phosphotungstic acid not a stain for polysaccharide. *Journal of Histochemistry and Cytochemistry* **18**, 455.

Gustafson, G. T. & Pihl, E. (1967). Staining of mast cell acid glycosaminoglycans in ultrathin sections by ruthenium red. *Nature, London* **216**, 697–698.

Khan, T. A. & Overton, J. (1970). Lanthanum staining of developing chick cartilage and reaggregating cartilage cells. *Journal of Cell Biology* **44**, 433–438.

Lesseps, R. J. (1967). The removal by phospholipase C of a layer of lanthanum-staining material external to the cell membrane in embryonic chick cells. *Journal of Cell Biology* **34**, 173–183.

Luft, J. H. (1966). Fine structure of capillary and endocapillary layer as revealed by ruthenium red. *Federation Proceedings. Federation of American Societies for Experimental Biology* **25**, 1773–1783.

Luft, J. H. (1968). Selective staining of acid mucopolysaccharides by ruthenium red. *Proceedings of the 26th Annual Meeting of the Electron Microscopic Society of America*, pp. 38–39 (Edited by C. J. Arceneaux), Baton Rouge: Claitor's Publishing Division.

Martinez-Palomo, A. (1970). The surface coats of animal cells. *International Review of Cytology* **29**, 29–75.

Pease, D. C. (1970). Phosphotungstic acid as a specific electron stain for complex carbohydrates. *Journal of Histochemistry and Cytochemistry* **18**, 455–458.

Rambourg, A. (1967*a*). An improved silver methenamine technique for the detection of periodic acid-reactive complex carbohydrates with the electron microscope. *Journal of Histochemistry and Cytochemistry* **15**, 409–412.

Rambourg, A. (1967*b*). Détection des glycoprotéines en microscopie électronique: coloration de la surface cellulaire et de l'appareil de Golgi par un melange acide chromique–phosphotungstique. *Compte rendu hebdomadaire des Séances de l'Académie des Sciences, Paris* D. **265**, 1426–1428.

Rambourg, A. & Leblond, C. P. (1967). Electron microscope observations on the carbohydrate-rich cell coat present at the surface of cells in the rat. *Journal of Cell Biology* **32**, 27–53.

Rambourg, A. (1969). Localisation ultrastructurale et nature de matériel coloré au niveau de la surface cellulaire par le mélange chromique–phosphotungstique. *Journal de Microscopie* **8**, 325–342.

Rambourg, A., Hernandez, W. & Leblond, C. P. (1969). Detection of complex carbohydrates in the Golgi apparatus of rat cells. *Journal of Cell Biology* **40**, 395–414.

Revel, J. P. (1964). A stain for the ultrastructural localization of acid mucopolysaccharides. *Journal de Microscopie* **3**, 535–544.

Revel, J. P. & Karnovsky, M. J. (1967). Hexagonal array of sunbunits in intercellular junctions of the mouse heart and liver. *Journal of Cell Biology* **33**, C7–C12.

Rothman, A. H. (1969). Alcian blue as an electron stain. *Experimental Cell Research* **58**, 117–179.

Scott, J. E. & Glick, D. (1971). The invalidity of 'Phosphotungstic acid as a specific electron stain for complex carbohydrates'. *Journal of Histochemistry and Cytochemistry* **19**, 63–64.

Seligman, A. M., Hanker, J. S., Wasserkrug, H., Dmochowski, H. & Katzoff, L. (1965). Histochemical demonstration of some oxidized macromolecules with thiocarbohydrazide (TCH) or thiosemicarbazide (TSC) and osmium tetroxide. *Journal of Histochemistry and Cytochemistry* **13**, 629–639.

Thiery, J. P. (1967). Mise en évidence des polysaccharides sur coupes fines en microscopie électronique. *Journal de Microscopie* **6**, 987–1018.

Tice, L. W. & Barrnett, R. J. (1962). Alcian blue staining for electron microscopy. *Journal of Histochemistry and Cytochemistry* **10**, 688–689.

Wetzel, M. G., Wetzel, B. K. & Spicer, S. S. (1966). Ultrastructural localization of acid mucosubstances in the mouse colon with iron containing stains. *Journal of Cell Biology* **30**, 299–315.

Enzyme digestion techniques

Aldridge, W. G. (1962). Studies on histochemical ribonuclease digestions. *Journal of Histochemistry and Cytochemistry* **10**, 682.

Alfert, M., Das, N. K. & Eastwood, J. M. (1962). Effects of RNAse on living cells and on cells fixed by different methods. *Journal of Histochemistry and Cytochemistry* **10**, 681–682.

Anderson, W. A. & André, J. (1968). The extraction of some cell components with pronase and pepsin from thin sections of tissue embedded in an Epon–Araldite mixture. *Journal de Microscopie* **7**, 343–354.

Anderson, W. A. & Ellis, R. A. (1965). Ultrastructure of *Trypanosoma lewisii*: flagellum, microtubules, and the kinetoplast. *Journal of Photozoology* **12**, 483–499.

Beaulaton, J. (1968). Sur l'action d'enzymes au niveau des pores nucleaires et d'autres structures de cellules sécrétrices prothoraciques incluses en Épon. *Zeitschrift für Zellforschung und microscopische Anatomie* **89**, 443–461.

Bernhard, W. (1966). Progress and limitations of ultrastructural cytochemistry carried out on ultrathin sections. *Journal of Histochemistry and Cytochemistry* **14**, 746–747.

Chopra, H. C., Shibley, G. P. & Walling, M. J. (1970). Electron microscopic cytochemistry of herpes simplex virus using enzyme extraction and autoradiography. *Journal de Microscopie* **9**, 167–176.

Douglas, W. H. J. (1970). Perchloric acid extraction of deoxyribonucleic acid from thin sections of Epon–Araldite-embedded material. *Journal of Histochemistry and Cytochemistry* **18**, 510–514.

Douglas, W. H. J., Ripley, R. C. & Ellis, A. (1970). Enzymatic digestion of desmosomes and hemidesmosome plaques performed on ultrathin sections. *Journal of Cell Biology* **44**, 211–215.

Granboulan, N. & Granboulan, P. (1965). Cytochimie ultrastructurale du nucleole. II. Études des sites de synthèse du RNA dans le nucleole et le noyau. *Experimental Cell Research* **38**, 604–619.

Leduc, E. H. & Bernhard, W. (1962). In *The Interpretation of Ultrastructure*, pp. 21–45 (Edited by R. J. C. Harris), New York and London: Academic Press.

Leduc, E. H. & Bernhard, W. (1967). Recent modifications of the glycol methacrylate embedding procedure. *Journal of Ultrastructure Research* **19**, 196–199.

Leduc, E. H. & Holt, S. J. (1965). Hydroxypropyl methacrylate, a new water-miscible embedding medium for electron microscopy. *Journal of Cell Biology* **26**, 137–155.

Leduc, E. H., Marinozzi, V. & Bernhard, W. (1963). The use of water-soluble glycol methacrylate in ultrastructural cytochemistry. *Journal of the Royal Microscopical Society* **81**, 119–130.

McGee-Russell, S. M. (1965). A ring and cup technique for handling thin sections for electron microscopy. *Journal of the Royal Microscopical Society* **85**, 91–96.

Marinozzi, V. (1964). Cytochemie ultrastructurale du nucléole—RNA et protéines intranucléolaires. *Journal of Ultrastructure Research* **10**, 433–456.

Marinozzi, V. & Bernhard, W. (1965). Présence dans le nucléole de deux types de ribonucleoprotéines morphologiquement distinctes. *Experimental Cell Research* **32**, 595–598.

Monneron, A. (1966). Utilisation de la pronase en cytochemie ultrastructurale. *Journal de Microscopie* **5**, 583–596.

Monneron, A. & Bernhard, W. (1966). Action de certaines enzymes sur des tissus inclus en Epon. *Journal de Microscopie* **5**, 697–714.

Moreno, G. & Vinzens, F. (1970). Comparaison en microscopie optique et électronique de l'action de la DNAse sur des cellules en couche monocellulaire. *Journal de Microscopie* **9**, 321–332.

Nass, S. & Nass, M. M. K. (1963). Intramitochondrial fibers with DNA characteristics. II. Enzymatic and other hydrolytic treatments. *Journal of Cell Biology* **17**, 613–629.

Pease, D. C. (1966*a*). The preservation of unfixed crytological detail by dehydration with 'inert agents'. *Journal of Ultrastructure Research* **14**, 356–378.

Pease, D. C. (1966*b*). Anhydrous ultrathin sectioning and staining for electron microscopy. *Journal of Ultrastructure Research* **14**, 379–390.

Rosenberg, M., Bartl, P. & Lesko, J. (1960). Water soluble methacrylate as an embedding medium for the preparation of ultrathin sections. *Journal of Ultrastructure Research* **4**, 298–303.

Smith, R. E. & Farquhar, M. G. (1965). Preparation of non-frozen sections for electron microscope cytochemistry. *Scientific Instrument News* **10**, 13–19.

Smith, R. E. (1970). Comparative evaluation of two instruments and procedures to cut nonfrozen sections. *Journal of Histochemistry and Cytochemistry,* **18**, 590–591.

Stäubli, W. (1963). A new embedding technique for electron microscopy, combining a water-soluble epoxy resin (Durcupan) with water insoluble Araldite. *Journal of Cell Biology* **16**, 197–201.

Weakley, B. S. (1969). Initial steps in the formation of cytoplasmic lamellae in the hamster oocyte and the identification of associated electron-dense particles. *Zietschrift für Zellforschung und microscopische Anatomie* **97**, 438–448.

Stains for 1 μm sections

Glauert, A. M. (1965). In *Techniques for Electron Microscopy*, 2nd edn, pp. 303–305 (Edited by D. H. Kay), Oxford: Blackwell Scientific Publications.

Lane, B. P. & Europa, D. L. (1965). Differential staining of ultrathin sections of Epon-embedded tissues for light microscopy. *Journal of Histochemistry and Cytochemistry* **13**, 579–582.

Stevens, A. R. (1966). In *Methods in Cell Physiology*, Vol. 2, pp. 274–285 (Edited by D. M. Prescott), New York and London: Academic Press.

Cryo-ultramicrotomy

Appleton, T. C. (1968). Ultrathin frozen sections for electron microscopy. Report published by LKB Instruments, Inc.

Bernhard, W. & Leduc, E. H. (1967). Ultrathin frozen sections. I. Methods and ultrastructural preservation. *Journal of Cell Biology* **34**, 757–771.

Christensen, A. K. (1967). A simple way to cut frozen thin sections of tissue at liquid nitrogen temperatures. *Anatomical Record* **157**, 227.

Christensen, A. K. (1968). Fresh-frozen thin sections cut, flattened and freeze-dried at a temperature of −75°C. *Journal of Cell Biology* **29**, 25A.

Fernández-Moran, H. (1952). Applications of the ultrathin freezing sectioning technique to the study of cell structures with the electron microscope. *Arkiv för Fysik* **4**, 471.

Hodson, S. & Marshall, J. (1969). Ultracryotomy: a technique for cutting ultrathin sections of unfixed frozen biological tissues for electron microscopy. *Journal of Microscopy* **91**, 105–117.

Leduc, E. H., Bernhard, W., Holt, S. J. & Tranzer, J. P. (1967). Ultrathin frozen sections. II. Demonstration of enzymic activity. *Journal of Cell Biology* **34**, 773–786.

The analytical electron microscope

Chandler, J. A. & Chou, C. K. (1970). The distribution of preservatives in wood and uptake by fungal hyphae, examined in the analytical electron microscope. *Proceedings of the International Congress of Electron Microscopy*, Vol. 3, pp. 493–494 (Edited by P. Favard), Paris: Societé Français de Microscopie Électronique.

Clarke, J. A., Salsbury, A. J. & Willoughby, D. A. (1970). Application of electron probe microanalysis and electron microscopy to the transfer of antigenic material. *Nature, Lond.* **227**, 69–71.

Cosslett, V. E. (1965). Possibilities and limitations of the differentiation of elements in the electron microscope. *Laboratory Investigation* **14**, 1009–1019.

Robertson, A. J. (1968). The electron probe microanalyzer and its applications in medicine. *Physics in Medicine and Biology* **13**, 505–522.

Schafer, P. W. & Chandler, J. A. (1970). Electron probe X-ray microanalysis of a normal centriole. *Science* **170**, 1204–1205.

9

Autoradiography with the Electron Microscope

One of the great limitations of the present day electron microscope is the impossibility of studying living material with it. However, the technique of autoradiography has proved invaluable in the investigation of tissue and cell dynamics at both the light and electron microscopic levels in recent years.

It should be stressed that autoradiography will yield information on where a material is synthesized, and, if properly timed studies are carried out, where the synthesized material goes. It will *not* disclose the presence of material which was synthesized before injection of the labelled precursor. It is not, therefore, a sophisticated cytochemical test for the presence or absence of a substance, but a device for studying the synthesis, turnover, and migration of materials. At the electron microscopic level it has helped to elucidate the roles of various organelles in the synthesis and distribution of cellular and extracellular substances.

The literature on the contributions of electron microscopic autoradiography to the dynamics of cell metabolism is too vast to cite in this book. Only references to basic papers and books relevant to the techniques described in the following pages will appear at the end of the chapter. For references to papers on particular aspects of cytodynamics see Baserga and Malamud (1969).

The Theory Behind the Technique

A radioactive isotope will decay with the release of energy to form another isotope. The amount of time it takes for one half the radioactive isotope to decay is termed its 'half-life', which is characteristic of the isotope in question. If the half-life is long compared with the duration of the autoradiographic experiment, the decay rate during the latter is effectively constant. In the process of decay, the radioactive substance emits radiation in the form of α, β or γ particles. If these particles are emitted sufficiently close to a photosensitive film the particles will interact with the silver halide granules so as to form latent images. Thus a record of the emission pattern can be trapped on the film.

Autoradiography is a technique based upon these facts. If one wishes to locate the site of synthesis of a particular material within a tissue or cell, one can select a chemical which is necessary for the synthesis of that material, and put a radioactive 'tag' or 'label' on it. The labelled chemical is then injected into an animal and enough time allowed to elapse for it to reach the tissue in question and to take part in the synthesis of the material to be located. The tissue is then removed from the animal, fixed, embedded and sectioned. The sections are placed in contact with a photographic emulsion and left in the dark long enough for the radioactive label to record its presence upon the emulsion. The emulsion can then be developed, fixed and studied in relation to the tissue.

Autoradiography in Light Microscopy

For light microscopy it has been found convenient to put a coat of liquid emulsion onto tissue sections which have been mounted on a glass slide. This ensures the necessary close contact between emulsion and specimen. Length of exposure is dependent upon several factors, including the half-life of the radioactive label and the amount of labelled material in the section. The latter, will, of course, vary with the metabolism of the particular tissue concerned and with the thickness of the section. If tritium is used as a label, exposure is usually one to four weeks. Specimen and emulsion are left in contact during development and fixation. The resulting preparation can be stained and viewed with the light microscope. The developed silver grains will be seen superimposed over those regions of the tissues where radioactive emission has occurred.

The procedures for light microscopy are fairly straight-forward and results are usually obtainable, after a little practice, within a few weeks.

Autoradiography in Electron Microscopy

The procedures for autoradiography in electron microscopy are both more difficult and more time-consuming than those for light microscopy. Tritium (a short-range β-emitter) is the only label which is widely used for electron microscopy, although ^{35}S has been used in certain cases. The half-life of tritium is $12\frac{1}{2}$ years. The time necessary for the tiny amount of material present in a 600 Å section to impart a useful amount of radiation to a photographic emulsion is usually between eight weeks and three months. There are so many techniques to be mastered, the variables to be considered so numerous, and the pitfalls so many that months of painstaking and frustrating work can be spent without any meaningful result. This is expensive in both time and materials.

Autoradiography at the ultrastructural level should not be undertaken without a thorough familiarity with the variables and pitfalls involved. It is not the purpose of this book to provide this background. The reader who is seriously going to engage in such work should immerse himself in the books and papers on theory and techniques listed at the end of the chapter.

If possible, a course of instruction in the techniques involved should be taken at some laboratory where this sort of work is being actively pursued.

Although tritium, being a short-range emitter, does not constitute a radiation hazard as long as it is handled with rubber gloves, its contamination hazard is considerable. This is especially true for such compounds as tritiated thymidine which if ingested can be incorporated into the DNA of cell nuclei throughout the body, with risk of serious tissue damage.

For the student who merely wishes to know in outline form what is involved, the following technique, which has provided useful results to the author, is described. It is based largely on the techniques of Salpeter and Bachmann (1964, 1965) and Bachmann and Salpeter (1965) as adapted by Rogers (1967).

Technique in Outline Form

1. Animals are injected intraperitoneally with a high-specific-activity radioactive tracer so as to provide a dose of 10 μCi/g body weight. The dose should be given in as concentrated form as possible to avoid the stress of a large volume of fluid in the peritoneal cavity. Controls are injected with the same volume of sterile saline. Tritiated compounds typically arrive in 10 ml rubber-capped glass vials containing 1 ml of concentrated radioactive tracer. Once the dosages required are calculated, the tracer can be diluted to the necessary strength by injecting sterile saline into the vial through the rubber cap. This of course involves increasing the pressure within the vial, so another sterile needle (without syringe attached) should be inserted through the cap at the same time to provide an exit for the replaced air. Otherwise, radioactive tracer will come bubbling out around the syringe in a dangerous manner.

2. Animals are killed at predetermined intervals depending upon the material to be studied. The literature should be consulted to

determine appropriate timings. It should be remembered that there may be large species differences in the speed with which a tracer is utilized metabolically, and trial and error may be necessary before the proper timing is arrived at.

3. The tissue is fixed in either glutaraldehyde, formaldehyde, or osmium tetroxide. Each has its disadvantages. Osmium tetroxide may accelerate fading of the latent image. Formaldehyde and glutaraldehyde may have a desensitizing effect on the emulsion. Many workers recommend formaldehyde as the lesser of the evils. Monneron (1969) prefers glutaraldehyde, as long as the tissue is washed in buffer for a minimum of two days with ten changes of fluid to prevent non-specific binding of label to the tissue. Glutaraldehyde appears to be a reasonable choice, since the ultrastructural preservation tends to be superior to that of formaldehyde. The particular properties and pitfalls of the fixative chosen must, of course, be kept in mind and allowed for.

4. The tissue is dehydrated and embedded. Once it is embedded, time is not a particularly critical factor because of the long half-life of tritium.

5. Sections are cut. Both one micron sections for light microscopy and ultrathin sections for electron microscopy are taken. The one micron sections are placed on a drop of water on slides which have previously been dipped in a gelatin solution (Rogers, 1967, p. 211) and allowed to dry down. The ultrathin sections are placed on collodion or Formvar-coated gold grids. (Gold grids are recommended since gold is less reactive than copper and hence less likely to cause chemical fogging of the emulsion. However, copper grids have been used by some workers, including the author, with good results.)

6. The sections are stained with uranyl acetate followed by lead citrate and then coated with a 200 Å thick film of carbon. This protects the emulsion from being fogged by the stain components. (Staining prior to exposure has been found to be generally more satisfactory than staining after exposure. The stain does not 'take' well through the emulsion layer, and there is danger of displacement of the label by the highly alkaline lead solution.)

7. The grids are attached to clean 1×3 inch glass microscope slides by a modification of the method of Hay and Revel (1963). Attachment is effected by use of double-sided Sellotape (1 inch wide) or Scotch double-sided masking Tape No. 400. These tapes have adhesive on both sides, one side being protected by paper. A hole slightly smaller than the grid is punched about one inch from the bottom of a 3 inch long strip of tape. The tape is then adhered to the slide and trimmed with a razor blade to an exact

fit. The protective paper is stripped from the top of the tape and the grid carefully centred over the hole. If the grid is positioned about one inch from the bottom of the slide the deposition of a uniform monolayer of emulsion over the grid is more likely to be attained when the slide is dipped.

8. The slides are taken in to a darkroom provided with an Ilford 902S safelight containing a 15 W bulb. Ilford L4 emulsion is melted in a water bath at 43°C and diluted, usually 1:8, with distilled water to which glycerol has been added to equal 1 per cent of the final volume. The glycerol prevents cracking of the emulsion during storage. Test slides are dipped in the emulsion and taken out of the darkroom to be checked. When dry, the emulsion layer should give a purple interference colour on that part of the slide where the grid would be positioned. This signifies that a monolayer of emulsion has been attained at that point.

9. The slides on which the grids are mounted are dipped into the emulsion (one quick dip each), drained vertically onto Kleenex for 30 seconds, and then placed vertically in a holder to dry. When the last slide has been dipped, the safelight is turned off and the slides allowed to dry in complete darkness for thirty minutes. They are then packed in light-tight slide boxes containing a drying agent (e.g. silica gel in porous plastic 'sausages') and placed in a refrigerator for exposure to take place.

One micron sections for light microscopy are processed in a similar manner except that the L4 emulsion is diluted 1:3, and after the initial 30 second vertical drainage onto Kleenex the slides are placed flat on a metal tray which has been pre-cooled by contact with containers filled with ice. They are then allowed to dry in complete darkness for 45 min before packing in light-tight boxes.

10. The time of exposure of the electron micrograph preparations is determined by reference to the time it takes the light microscopic preparations to yield satisfactory results. Light slides are checked at intervals, usually starting after 7 days of exposure, to determine when developed grains are present. The slides may be developed for 3 to 5 min in any of the common developers, e.g. PQ universal. They may then be stained at room temperature in a 1:1 mixture of 1 per cent methylene blue and 1 per cent borax in a coplin jar for 15 min to 1 hour (depending on the particular batch of methylene blue). When adequate labelling has occurred in the light slides, the time this has taken is multiplied by ten to get a rough idea when the electron microscope preparations will be ready. This is, it is stressed, only a rough idea. It is wise to start checking the electron microscope preparations after about six weeks.

Although Ilford L4 may have a useful life of about 6 months, an exposure time of more than 2 months duration runs the risk of increased background due to the age of the emulsion.

11. The preparations for the electron microscope are developed in Microdol X developer for 3 to 4 min. This is a fine-grain developer which will yield a developed grain size of 0·12 to 0·16 μm. Thus a resolution of approximately the same dimension will be obtained. (This developer cannot, of course, be used for the light preparations since the developed grain size falls below the resolution of the light microscope.)

After development the grids are rinsed in two changes of distilled water, fixed for not more than one minute in Hypam Rapid Fixative, and rinsed for a total of 15 min in three changes of distilled water. The grids are then removed from the tape, allowed to dry, and placed in gelatine capsules or other containers until they are viewed. Removal from the tape must be done with great care under a dissecting microscope to avoid distorting or tearing the grids. If removal is done before the tape has become dry after rinsing it is slightly easier.

The grids are now ready for inspection. One will already have some idea of what to expect from the light microscopic preparations. In fact, most authorities recommend doing preliminary light microscopic autoradiography before attempting any work at the electron microscope level. However, since better autoradiographic resolution at the light level can be attained by using one micron sections from tissues embedded in Araldite, and since the same L4 emulsion can be used for both light and electron microscope preparations, it seems more economical of both time and materials to prepare specimens for electron microscopy at the same time from the same blocks. This can, at the very least, serve as a 'practice run' for the demanding electron autoradiographic preparative techniques, and at best may provide usable specimens for study.

Rogers (1967) recommends Bachmann and Salpeter's technique of mounting ultrathin sections upon collodion-covered glass microscope slides. Then after exposure and development the collodion film is floated from the slide onto a water surface and the grids dropped upon the area containing the sections. This technique has worked well in many laboratories. It is thus largely a matter of personal preference which system one chooses. Both have their inherent difficulties; both are capable of yielding good results in practised hands.

Radioactively labelled materials are obtainable (by personnel in laboratories who have been properly certified under the Radioactive Substances Act) from The Radiochemical Centre, Amersham, England. They will send a catalogue, brochures and information sheets on request.

REFERENCES

Official publications

Department of Employment and Productivity (1968). *Code of Practice for the Protection of Persons Exposed to Ionizing Radiations in Research and Teaching*, London: Her Majesty's Stationery Office.

Committee of Vice Chancellors and Principals (1966). *Radiological Proceedures in Universities* (Handbook for use in universities to supplement the Ministry of Labour's Code of Practice for the Protection of Persons Exposed to Ionizing Radiations in Research and Teaching), London: Association of Commonwealth Universities.

General references

Bachman, L. & Salpeter, M. M. (1965). Autoradiography with the electron microscope. A quantitative evaluation. *Laboratory Investigation* **14**, 1041–1053.

Baserga, R. & Malamud, D. (1969). *Autoradiography. Techniques and Applications*, New York: Hoeber.

Cleaver, J. E. (1967). *Thymidine Metabolism and Cell Kinetics*, Amsterdam: North Holland Publishing.

Evans, E. A. (1966). *Tritium and its Compounds*, London: Butterworths.

Feinendegen, L. E. (1967). *Tritium-Labeled Molecules in Biology and Medicine*, New York and London: Academic Press.

Gude, W. D. (1968). *Autoradiographic Techniques—Localization of Radioisotopes in Biological Material*, Englewood Cliffs, N.J.: Prentice-Hall.

Hay, E. D. & Revel, J. P. (1963). The fine structure of the DNP component of the nucleus. An electron microscopic study utilizing autoradiography to localize DNA synthesis. *Journal of Cell Biology* **16**, 29–30.

Monneron, A. & Moulé, Y. (1969). Critical evaluation of specificity in microscopical radioautography in animal tissues. *Experimental Cell Research* **56**, 179–193.

Rogers, A. W. (1967). *Techniques of Autoradiography*, Amsterdam, London and New York: Elsevier.

Salpeter, M. M. (1966). *In Methods of Cell Physiology*, Vol. 2, pp. 229–253 (Edited by D. M. Prescott), New York and London: Academic Press.

Salpeter, M. M. & Bachmann, L. (1964). Autoradiography with the electron microscope. *Journal of Cell Biology* **22**, 469–477.

Salpeter, M. M. & Bachmann (1965). In *The Use of Autoradiography in the Investigation of Protein Synthesis* (Edited by R. J. C. Harris), New York: Academic Press.

Sjöstrand, F. S. (1967). *Electron Microscopy of Cells and Tissues*, Vol. 1, Chap. 17, New York and London: Academic Press.

Stevens, A. R. (1966). In *Methods in Cell Physiology*, Vol. 2, pp. 255–310, New York and London: Academic Press.

Williams, M. A. (1969). In *Advances in Optical and Electron Microscopy*, Vol. 3, pp. 219–279 (Edited by R. Barer and V. E. Cosslett), London and New York: Academic Press.

10

A Brief Synopsis of Other Techniques for the Electron Microscope

THE FREEZING TECHNIQUES

In the chapter on the processing of tissues it was made clear that tissue is changed in radical ways by the processes to which it is subjected after removal from the animal. Most particularly, the reactions involved in chemical fixation are of prime importance in determining what the tissue will ultimately look like when viewed under the electron microscope. Different fixatives allow materials to be leached out of the cells to varying degrees. Much metabolically important material is removed altogether. The molecular configurations of the cell structure that is left may be completely altered by the action of the fixative.

How closely does the picture of a chemically fixed cell correspond to reality? Obviously, in order to determine this, some independent method must be used which does not involve chemical fixation.

FREEZE DRYING

One such method, freeze drying, involves rapid freezing of the tissue, followed by desiccation. The freeze drying method, which had already been used in other fields, was adapted for electron microscopy by Sjöstrand and Baker (1958), and modified by Hanzon and Hermodsson (1960) and Seno and Yashizawa (1960), among others.

If a small block of tissue is immersed in propane, isopentane or Freon cooled to the temperature of liquid nitrogen (196 °C), freezing is nearly instantaneous. The tissue is stabilized by the cold and all metabolic processes are halted in far less time than it takes a chemical fixative to penetrate the block. If freezing is accomplished fast enough, ice crystals do not form in the tissue. This is because a regularly repeating three-dimensional crystal lattice does not have time to form before the molecules get too sluggish from the cold to link up with one another. The water passes into a vitreous (non-crystalline) state in which the molecular pattern is random. This does not disrupt the structure of the cell as ice crystal formation would do. The vitrified water is then sublimed off by placing the tissue in a high vacuum (10^{-6} to 10^{-7} mmHg). After this drying process, the tissue is infiltrated with embedding medium. This is usually methacrylate, although both Araldite and Vestopal W have been used. The methacrylate is immersed in liquid air to solidify it, and the tissue is then placed upon the solidified embedding material. As the temperature is raised the methacrylate melts and penetrates the tissue. The methacrylate is then polymerized by ultraviolet radiation at $-10°$C. The embedded tissue is brought to room temperature and may then be sectioned and stained in the usual way. Thus both chemical fixation and subjection to dehydrating solvents are avoided.

Several difficulties are encountered in the freeze-drying technique. One is that it is extremely difficult in practice to freeze the tissue quickly enough to avoid the formation of ice crystals. Pure water is extracted from the cell protoplasm to form the ice crystals, thereby concentrating the non-aqueous material of the cytoplasm. This causes damage to organelles through osmotic effects, and also through pH changes (Bullivant, 1970). The ice crystals may cause displacement or even rupture of organelles. 'Ghosts' of these crystals are seen as large, transparent areas in the final electron micrographs. A great deal of time is spent looking for areas of tissue which are sufficiently free from ice crystal artefacts to yield meaningful information.

The ice crystal problem has been alleviated somewhat by placing the tissue in 20 to 30 per cent glycerol before freezing. Glycerol and water are completely miscible in all proportions. The glycerol molecules bond strongly to the water molecules by hydrogen bonds. This separates the water molecules from one another and prevents the formation of ice crystals. Glycerol is thought to be sufficiently inert to provide this protection without reacting chemically with any of the other cell constituents.

Another difficulty is that the removal of water by sublimation leaves the tissue liable to collapse, especially during infiltration with the embedding material. As the embedding material 'sets', polymerization damage may occur. Also, since the embedding medium itself is not chemically inert, it may react with tissue constituents, altering their structures. Lipid appears to be extracted from the tissue both by the methacrylate and by the acetone solution used in the cutting bath; thus membranes appear as negative images. Very little protein is lost; thus the density of the cytoplasmic and organelle matrices is high. Contrast in the final electron micrograph is, as a consequence, rather poor.

Despite these problems, freeze drying has provided a comforting and rather unexpected confirmation that cell ultrastructure as seen after chemical fixation does bear some relation to reality, at least in the medium resolution range. The organelles of freeze-dried tissue, although they differ somewhat in density and in staining properties, have in general the same size, distribution and ultrastructure as that seen after chemical fixation.

At the high resolution macromolecular level, however, the technique has a limited usefulness. As Bullivant (1970) points out,

> 'The structure of proteins, nucleic acids and lipids depends on their interaction with the surrounding layers of water, and the configuration of such large molecules is changed when water is replaced by a nonpolar embedding medium prior to thin sectioning.'

Freeze Substitution

Freeze substitution is a variant of freeze drying, and is easier to perform. It was introduced by Fernández-Morán (1959, 1960) and Bullivant (1960). The initial freezing is similar to that used in the freeze-drying technique, but the vitrified water is removed by immersing the frozen tissue in acetone, alcohol or some other dehydrant before embedding. At these low temperatures acetone or alcohol have little effect on tissue structure. If desired, however, the tissue may be chemically fixed after freezing. This may be done by exposure to fixative vapour, or by dissolving the fixative in the dehydrant. Fixation, of course, partly negates the original purpose of the freeze-drying method, namely to avoid the artefacts caused by the use of chemicals. There are, however, several advantages in the freeze substitution technique. Damage caused by infiltration of the embedding material is reduced. Since there is no free water in the frozen tissue, water-soluble substances are not extracted during processing. Also, if osmium tetroxide is used as a fixative, lipid-containing membranes are retained and their structure can be compared with that of membranes preserved by other methods.

In general, the ultrastructure of the cell and its organelles as seen after freeze-substitution is comparable to that after routine fixing, embedding and sectioning. Electron density of many components is greater, as more material is preserved, and as in freeze-drying, contrast is poor. Membranes in the mitochondrial cristae appear to fuse with one another (Malhotra, 1968) and the dimension of the periodicity myelin is different, but otherwise membrane structure is not greatly altered at the medium resolution level.

For an excellent review of freeze drying and freeze substitution techniques, see Bullivant (1970).

Ultrathin Frozen Sections (Cryo-ultramicrotomy)

This technique was described in Chapter 8, and pertinent references will be found at the end of that chapter.

Freeze Etching

This elegant technique was introduced by Steer (1957) and developed further by several workers, including Moor, Muhlethaler, Waldner and Frey-Wyssling (1961) and Bullivant and Ames (1966). More recent modifications are described by Bishop (1969), Moor (1969) and Bullivant (1970). The technique was slow to be adopted because of the extreme difficulty of preparing the specimens. However, it is now coming into increasing use as the equipment and techniques involved have been refined.

Freeze etching is a purely physical method and entirely avoids the chemical and structural alterations in the tissue which occur during chemical fixation, dehydration, embedding and sectioning. Since no substances are washed out of the tissue, no shrinkage can occur.

The degree to which the photographs obtained with this technique may correspond to the actual structure of living tissue may be inferred from the fact that living yeast cells whose surfaces have been studied in this manner have been able to resume apparently normal functioning and to reproduce themselves (Moor, 1966).

The tissue is soaked in glycerol and frozen in liquid Freon maintained at $-150\,^{\circ}C$ by liquid nitrogen. Next the tissue is placed in liquid nitrogen at $-196\,^{\circ}C$. After removal from the liquid nitrogen, it is placed on a cold table and fractured with a razor blade so that the cells are split across (Fig. 15). The fractured surface is exposed to a high vacuum (10^{-6} to 10^{-7} mmHg). As the vitrified water is sublimed away, the cut surface is 'etched' so that the organelles stand out in relief. A coat of platinum–carbon is evaporated onto this etched surface from a platinum–carbon simultaneous mixture electrode system. Thus a replica of the surface is formed. (For details of replica formation, see pp. 164–166.) The platinum–carbon

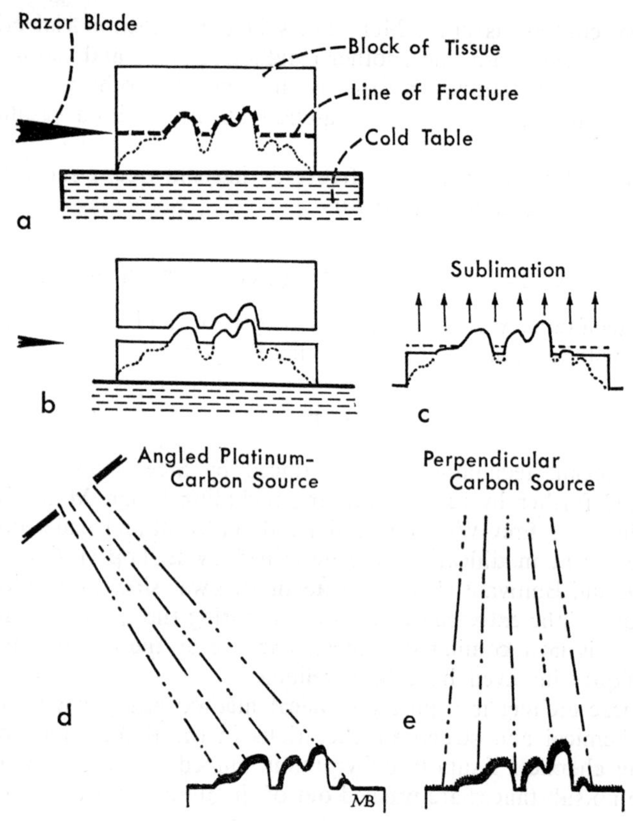

Fig. 15. Freeze-etching procedures. (*a*) Razor blade is used to fracture frozen tissues. (*b*) Fracture splits tissue along natural surfaces. (*c*) Sublimation of water etches surface. (*d*) Replica formation: shadowing with angled platinum-carbon source. (*e*) Strengthening replica by evaporation from perpendicular carbon source.

evaporation is done at an angle so that the structures proud to the etched surface are shadowed, thus increasing the contrast of the replica. The replica is strengthened by perpendicular evaporation of a layer of carbon onto it. The tissue is then allowed to warm up in distilled water and the replica floats off. Traces of organic material remaining on the replica are removed with trypsin followed by sulphuric acid, or strong alkali. After thorough rinsing in distilled water the replica is picked up on a grid and is ready to be viewed in the electron microscope.

The most fascinating and informative aspect of this technique is that the fractured surface reveals details which are obtainable by no other method. In following the line of least resistance, the fracture tends to proceed along natural surfaces within the cells. Such natural surfaces include the membranes of such organelles as the nucleus, mitochondria, golgi apparatus and vesicles. Thus the surfaces of these organelles are exposed and may be studied in detail. For instance, the distribution of nuclear pores over an entire half of a nucleus may be observed. In this way it has been found that pore distribution varies greatly from one cell type to another —i.e. the pores may delineate hexagons in one cell type, circles in another, parallel lines in another (Northcote and Lewis, 1968).

Cross-fracture of the organelles may also occur, depending upon the method and conditions of fracture, and fascinating three-dimensional views of organelle interiors may be obtained.

In broad terms, the picture of cellular ultrastructure provided by chemical fixation and preparation of thin sections has been shown by freeze etching to be essentially correct. However, interpretation of the surface views obtained by this method is difficult. When the fracture follows the membrane of an organelle, does it follow the external surface, the internal surface, or split the membrane sheet apart? Can all three of these possibilities occur depending upon the type of organelle and the conditions of fracture? Moor (1966, 1969) has studied complementary replicas taken from the two surfaces produced from single fractures and concludes that the fracture occurs along the membrane surface. Branton (1966, 1969) and Da Silva and Branton (1970) studying the same sort of material conclude that the membrane is split down the middle on fracture, so that the surface exposed is presumably composed of the non-polar (hydrophobic) ends of phospholipid molecules. They suggest that the hydrophobic bonds are weakened by freezing.

On many membrane surfaces thus exposed, small particles, usually about 100 Å in diameter, or small pockets complementary to these particles, are visible. The number and distribution of the particles vary from one organelle to another, but the differences are consistent. This suggests that the particles reflect some real property of the membranes. Da Silva and Branton (1970) suggest that the 100 Å particles (which are probably closer to 40 to 50 Å in diameter when the enlargement caused by shadowing is allowed for) represent either particles of protein intercalated in the phospholipid bilayer, or micellar aggregations of the phospholipid itself. If this interpretation is correct, this view of the inner membrane structure is one which has been seen by no other technique. It suggests intriguing possibilities for the elucidation of membrane structure.

The resolution obtainable with the freeze-etching technique depends upon the heavy metal used for producing the replica. With platinum–carbon replication the resolution is limited to 20 to 30 Å. However, work with tantalum and certain other metals indicate that the resolution attainable may soon reach 10 Å (Moor, 1969). Ten ångström resolution is, of course, far better than is routinely obtained using ultrathin sections.

For recent reviews of freeze-etching, see Moor (1969) and Branton (1969).

In summary, then, it may be said that the freezing techniques have provided a valuable confirmation of the results obtained by classical electron microscopic techniques. The difficulties involved in freeze drying and freeze substitution have prevented their widespread use. However, the technical advances made in the field of freeze etching have resulted in increasing numbers of workers using the method. Highly informative results should be forthcoming as histochemical and cytochemical methods are applied to this medium.

THE STUDY OF SURFACES BY ELECTRON MICROSCOPY

Thin Film Replicas and Shadow-Casting

In order for the electron beam to penetrate a specimen and yield a meaningful image, the specimen must ordinarily be thinner than 1200 Å. Before ultrathin sectioning techniques were developed in the early 1950s, thin specimens were obtained by making thin-film replicas of the surfaces of material under investigation. Replicas are also an integral part of the freeze-etching technique, as described in the preceding section.

A number of techniques for replica making have been used with biological material. Two of the most useful are (1) replicas made by forming a thin film of collodion or Formvar on the specimen surface, and (2) replicas formed by depositing a thin film of material (usually metal or carbon) onto the specimen by evaporation.

In order to form a replica by evaporation, the specimen and the material to be evaporated are placed in a vacuum chamber and the pressure reduced to 10^{-5} mmHg. This prevents interference with the evaporating process by air molecules. The substance to be evaporated is heated by passing a current through a V-shaped tungsten wire. The tip of the V (which becomes the hottest part of the wire) is in contact with the substance. Both silicon and carbon, as well as certain metals, have been used for making thin film replicas. Carbon is especially useful, since it evaporates readily and is so inert that the tissue can be dissolved away from the replica without damaging it.

If a replica film is of uniform thickness and composition it will, of course, lack contrast when viewed in the electron microscope. Contrast can be produced in a replica by evaporating electron-dense material, such as a heavy metal, onto the replica at an oblique angle. The metal atoms will pile up on the side of surface contours nearest the evaporation source, and be reduced in amount on the side furthest from it (Fig. 15*d*). This is analogous to an object in bright sunlight being illuminated on the side nearest the sun, and casting a shadow on the side away from the sun. The process is therefore known as shadow-casting. A positive print made from a negative of a shadow-cast surface will show dark masses on the side of contours nearest the evaporating source, and pale shadows on the other. Therefore, negative prints are usually made in order for the shadowed material to appear more consistent with common experience.

If the angle of evaporation (θ) is known and the length (l) of the shadow determined by measuring it on the negative print and dividing by the total magnification, the height (h) of the object may be calculated from the formula

$$h = l \cdot \tan\theta$$

Replication and shadowing can be done in one step if platinum and carbon are simultaneously evaporated at an angle onto the specimen from a platinum–carbon simultaneous mixture electrode.

The resolution of fine detail obtainable with the replica technique depends upon the material used in making the replica. Collodion and Formvar can only be used if high resolution is not required. Some metals, e.g. gold, tend to condense in coarse grains after evaporation. Lighter metals will not produce sufficient contrast in thin films. It has been found (Bradley, 1959) that the simultaneous evaporation of platinum and carbon prevents large grains of platinum from building up. This combination will give very sharp shadows and a resolution of approximately 20 to 30 Å.

In studying certain biological materials it is unnecessary to dissolve away the specimen from the replica. Small particulate matter such as cell organelles and fibres isolated from fragmented tissues, viruses, bacteria and macromolecules such as globular protein and nucleic acids may be dried down onto a supporting film, shadowed, and studied without further processing. This, in effect, might be called a form of negative staining, the 'stain' in this case being the evaporated material.

For more detailed information on replica formation and shadowing, see Pease (1964) and Bradley (1965).

The Scanning Electron Microscope

The scanning electron microscope represents a major advance over the replica technique for the study of surfaces. The first scanning electron microscope was built in Germany in 1938, but it was many years before improvements in design made commercial production feasible. The first commercial instrument was put on the market in 1965 by the Cambridge Instrument Company, using a design based on the work of Professor C. W. Oatley and coworkers at Cambridge University (Oatley, Nixon and Pease, 1965; Oatley, 1966).

In scanning electron microscopy, an ultrafine electron probe scans the surface of the specimen in a square raster, and the detected information is imaged on a cathode ray tube carrying a long-persistence phosphor. A second, high resolution, short-persistence cathode ray tube is used when taking photographs of the specimen.

The type of information which the probe detects varies with the mode of operation of the machine. Any detectable phenomena which result from the interaction of the electron beam and the specimen can be so manipulated electronically as to produce a visual image. Primary electrons scattered from the specimen surface, secondary electrons knocked from atoms in the specimen, and absorbed primary electrons can all contribute information about the specimen. Distribution of elements within a specimen can be detected by collecting the back-scattered electrons from the specimen surface layers, since more electrons are back-scattered from elements of higher atomic number than from elements of lower atomic number. Luminescence caused by excitation of specimen components by the electron beam can be detected, but as yet has no important biological applications.

The scanning electron microscope utilizes a lower operating voltage than does the conventional transmission electron microscope, the range being usually 1 to 30 kV. The useful magnification range runs from 10 to 50,000 times, thus providing a bridge between the light and transmission electron microscope ranges.

With the appropriate attachments, scanning microscopes can also be used for transmission electron microscopy. Also available are special scanning attachments which can be used with a conventional transmission electron microscope. The conversion from the scanning mode to the transmission mode and vice versa takes only a few minutes, so that the same specimen may be studied by both methods within a short period of time and without moving the specimen.

The emissive mode of scanning operation is most commonly used for the study of biological specimens. It detects low energy secondary electrons emitted from the specimen surface. Using this mode, surface detail is resolvable down to 100 Å under optimal operating conditions. Because of the great depth of focus of the scanning microscope (which is hundreds of times greater than that of the light microscope) the surface views obtained have a very nearly three-dimensional appearance (see Plate 5). Details in crevices and holes and behind protruding surfaces are visible, since the scanning electron beam produces detectable secondary electrons from the walls of these structures (Heywood, 1969). It may be difficult to judge, however, whether a structure protrudes from the surface or dives down into the surface. This problem is resolved by the technique of stereophotography.

In stereophotography, two photographs of a specimen are taken at an angular separation of at least 5 to 6°. The two photographs are then projected from a special stereo-projector onto a single screen. If the proper projection angles are used and special viewing glasses employed, a striking three-dimensional picture of the specimen surface is obtained.

In biological work the scanning electron microscope has so far found its greatest usefulness in the study of plant materials. These materials need little or no special preparation before they are studied. In botany, surface features of seeds, fruits, pollen, fungal spores, etc. are of great importance in the areas of taxonomy, morphology and diagnosis (Heywood, 1969). The pulp and paper industry has made extensive studies of the microstructure of wood and has gained much useful information on both normal morphology and on the effects of chemical and physical processes on wood fibres. Other particularly fascinating work is being done in paleobotany and the investigation of microfossils.

The study of soft animal tissues with the scanning electron microscope presents special problems. Special preparative procedures must be applied before soft tissues can be investigated. First, water must be removed by some means which does not collapse the remaining structure. This may be accomplished either by freeze-drying or by fixation followed by alcoholic dehydration. The tissue is then secured onto a specimen holder and coated with a thin layer of gold–palladium alloy. This metal layer makes the specimen surface conductive and prevents the building up of charges, which would otherwise adversely affect resolution. The specimen is then ready for viewing.

The epithelial, mesothelial and endothelial surfaces of tissues may be fairly easily prepared for study, as can isolated cells, tissue cultures and smears. Selective digestion, such as the removal of collagen from interstitial areas may be helpful in some cases. If internal structure is to be viewed, however, the fixed tissue must be dissected, torn apart or fractured in order to expose internal surfaces. Sections of embedded material are not suitable for scanning; their smooth, evenly cut surface reveals no meaningful information (Boyde and Wood, 1969).

The elaboration of techniques for the scanning electron microscope is still in its infancy. As the methods improve, there will no doubt be a great extension of the areas in which this versatile research tool can be usefully applied.

HIGH VOLTAGE ELECTRON MICROSCOPY

Electron microscopes employing an accelerating potential of one million to three million volts have been developed and are now operating in several countries, notably France, Japan, Great Britain and the United States.

The cost of producing such a microscope (roughly £200,000) quite effectively keeps the number of machines to a minimum. In spite of this, the number now in use exceeds twenty, and more are under construction. The usual practice is for one high voltage machine to serve scientists over a large area. In Britain the AEI EM7 one million volt microscope located at the Atomic Energy Research Establishment, Harwell, Berkshire, is available for use by experienced 100 kV microscopists at a nominal fee.

The Rationale for High Voltage Electron Microscopes

As the accelerating voltage propelling an electron beam increases, the energy of the electrons in the beam increases. As electron energy increases it becomes more difficult to stop or deflect individual electrons. Thus with higher voltages, thicker specimens can be penetrated by the beam.

As the accelerating voltage increases, the wavelength associated with the electron beam goes down. The net effect of this decrease in wavelength in terms of resolving power is not momentous: the theoretical resolving power of the conventional 100 kV instrument is 2·7 Å due to the various considerations described in Chapter 1, and that of the one million volt machine is 1·1 Å (Dupouy, 1968).

However, since the scattering of electrons by a specimen of given thickness lessens as acceleration increases, chromatic aberration produced by this scattering is reduced. This can be expected to improve resolution considerably when studying sections of biological material.

Before the advent of thin sectioning techniques there was considerable interest in producing an extra-high voltage electron microscope in the hope of penetrating thick specimens. Several prototype instruments were built in the late 1940s. However, since the immediate problem of viewing internal structure was solved by the thin sectioning technique, commercial interests were directed mainly to perfecting the 50 to 100 kV instruments already in use.

The conventional 50 to 100 kV transmission electron microscope used in conjunction with ultrathin sections has yielded such a wealth of information that biologists have been kept busy for the past twenty years recording and interpreting it. Consequently, pressure for high voltage machines has come mainly from the materials industry, and it may be that their greatest contribution will be in this field. Some workers in the life sciences contend that little meaningful biological information will come from ultrahigh voltage microscopy. Others are confident that there is great scope for high voltage microscopy in biology and medicine (Cosslett, 1969).

Biological Applications of High Voltage Electron Microscopy

At present, work on biological materials is proceeding very slowly, and is doing so in two directions which will be briefly described below.

1. *The Microscopy of Living Material*

Dupouy and colleagues in France have designed a special specimen chamber in which bacteria and other living material can be kept in a moist atmosphere. The chamber has two windows of thin plastic through which the electron beam can pass. Bacteria have been studied and photographed in this chamber and then removed and placed in culture media. They are said to have withstood the irradiation by the electron beam well enough to remain alive and to reproduce themselves (Dupouy, Perrier and Durrieu, 1960; Dupouy, 1968). The photographs of living material obtained to date are not particularly informative, but the potential of the method is still to be realized. Plate 6 is a micrograph of a bacterium viewed with an electron beam of one million volts.

2. *Three-dimensional Stereoelectron Microscopy*

Information on the three-dimensional appearance of tissues, cells and organelles is extremely difficult to obtain using ultrathin sections. As pointed out in Chapter 7, serial sectioning and photographing one cubic millimetre of tissue would require 4×10^8 photographic exposures. The time which would be consumed in constructing a three-dimensional model from these photographs boggles the imagination.

A more direct approach is stereophotography of the thicker sections which can be studied with ultrahigh voltage electron microscopes (Porter and Hama, 1968; Cosslett, 1969; Hama and Porter, 1969; Nagata, Hama and Porter, 1969; Ris, 1969; Hama, 1970). Stereo methods are essential here in order to make sense out of the otherwise meaningless jumble of overlapping structures seen in the thick sections. The extra penetrating power of the high voltage beam, coupled with specimen tilting and stereo techniques (see below) should yield an increasing amount of valuable information. The techniques are beset with difficulties, but these do not appear to be insuperable.

TILTING SPECIMEN STAGES

If components in the specimen, for example membranes, happen to be aligned exactly on the axis of the electron beam (i.e. so that they do not pass obliquely through the section) contrast and resolution are improved because the underlying structure reinforces that at the section surface rather than detracting from it. Superposition effects can thus be minimized by proper orientation of suitable specimens. In this way, resolutions of better than 30 Å have been attained with stained sections as thick as 3000 Å, using an accelerating voltage of only 100 kV (Wachtel *et al.*, 1966).

Obviously if the specimen can be tilted so as to ensure alignment of specimen components parallel to the beam it is a tremendous advantage. The newer transmission electron microscopes, including the high voltage machines, can be fitted with special specimen stages in which the specimen may be tilted at angles up to $\pm 45°$ in two mutually perpendicular directions and in some cases rotate through 360° as well. By tilting a section it is possible to bring structures which pass obliquely through the section, such as membranes or filaments, into register with the electron beam. Membranes which give a diffuse image when oblique to the beam give a clear indication of their transverse structure when lying on the same axis as the beam (see Fig. 13, Chapter 7).

Tilting the specimen also allows photographs of structures lying within a thin section (or a thicker one in case of high voltage electron microscopy) to be photographed from many points of view. Small structures such as ribosomes and glycogen particles may be completely enclosed within one thin section. By tilting the section a great deal of information about the three-dimensional structure of such bodies may be obtained. By taking photographs of the same structure from two different angles, a 'stereo-pair' of photographs can be produced which, seen through an appropriate viewer, will give a three-dimensional rendering of the specimen. According to Gray and Willis (1968), micrographs taken with a tilt angle difference of up to 20° can be fused as a stereo-pair. With a tilt angle difference of more than 20° the pictures become too dissimilar to fuse.

PHOTOGRAPHING REINFORCEMENT OF SPECIMEN DETAIL

The Markham Rotation Technique

Markham, Frey and Hills (1963) introduced a photographic rotation technique for improving the resolution of electron micrographs of certain types of material. If an object has a natural radial symmetry, this symmetry will be reinforced if the object is photographed and then rotated an appropriate number of degrees and rephotographed on the same plate. As the symmetry is reinforced, details which have no symmetry are blurred out.

The method has been very helpful in elucidating the structure of certain viruses. However, it is a technique in which it is easy to obtain spurious results, and many papers have been written on material which is inherently incapable, through lack of the required natural symmetry, to give meaningful results with this method. (For review and discussion see Friedman, 1970.) Nevertheless, when properly applied, the Markham rotation technique is a very useful one.

The Electron Micrograph Optical Diffractometer

Near the limit of resolution of the electron microscope, problems arise in differentiating random noise and phase contrast effects from the true structure of the specimen. It has been found that optical diffraction methods can be applied to high resolution electron micrographs to sort out structure from noise. The result has been the development of the electron micrograph optical diffractometer based on designs by Professor R. Markham and Dr R. W. Horne. As in the Markham rotation technique, the specimen must possess some regularity of structure such as periodicity or symmetry. The effects of astigmatism, specimen drift and focusing can also be identified by this method. For details of the method and references for further reading see Markham (1968).

REFERENCES

Freezing techniques

Bishop, W. R. (1969). Methodology of rapid freeze cleavage replication of mammalian tissues and cells. *Balzers High Vacuum Report* No. **24**, 1–19.

Branton, D. (1966). Fracture faces of frozen membranes. *Proceedings of the National Academy of Sciences of the U.S.A.* **55**, 1048–1056.

Branton, D. (1969). Membrane Structure. *Annual Review of Plant Physiology* **20**, 209–238.

Bullivant, S. (1960). The staining of thin sections of mouse pancreas prepared by the Fernandez-Moran helium II freeze-substitution method. *Journal of Biophysical and Biochemical Cytology* **8**, 639–647.

Bullivant, S. (1970). In *Some Techniques in Electron Microscopy*, pp. 101–146 (Edited by D. F. Parsons), New York and London: Academic Press.

Bullivant, S. & Ames, A. (1966). A simple freeze-fracture replication method for electron microscopy. *Journal of Cell Biology* **29**, 435–447.

Fernández-Morán, H. (1959). Electron microscopy of retinal rods in relation to localization of rhodopsin. *Science*, New York **129**, 1284–1285.

Fernández-Morán, H. (1960). Low temperature preparation techniques for electron microscopy of biological specimens based on rapid freezing with liquid helium II. *Annals of the New York Academy of Sciences* **85**, 689–713.

Hanzon, V. & Hermodsson, L. H. (1960). Freeze-drying of tissues for light and electron microscopy. *Journal of Ultrastructure Research* **4**, 332–348.

Malhotra, S. K. (1968). In *Cell Structure and its Interpretation*, pp. 11–21 (Edited by S. M. McGee-Russell and K. F. A. Ross), London: Edward Arnold.

Moor, H. (1965). Freeze-etching. *Balzers High Vacuum Report* **2**, 1–23, Balzers, Principality of Liechtenstein: Balzers, A. G.

Moor, H. (1966). The performance of freeze-etching and the interpretation of results concerning the surface structure of membranes and the fine structure of microtubules and spindle fibres. *Balzers High Vacuum Report* **9**, 1–11.

Moor, H. (1969). Freeze-etching. *International Review of Cytology* **25**, 391–412.

Moor. H., Muhlethaler, K., Waldner, H. & Frey-Wyssling, A. (1961). A new freezing ultramocrotome. *Journal of Biophysical and Biochemical Cytology* **10**, 1–13.

Northcote, D. H. & Lewis, D. R. (1968). Freeze-etched surfaces of membranes and organelles in the cells of pea root tips. *Journal of Cell Science* **3**, 199–206.

Pease, D. C. (1967). Eutectic ethylene glycol and pure propylene glycol as substituting media for the dehydration of frozen tissue. *Journal of Ultrastructure Research* **21**, 75–97.

Rebhun, L. I. (1965). Freeze-substitution: fine structure as a function of water concentration in cells. *Federation Proceedings. Federation of*

American Societies for Experimental Biology **24**, Suppl. 15, S-217 to S-232.

Seno, S. & Yashizawa, K. (1960). Electron microscopic observations on frozen-dried cells. *Journal of Biophysical and Biochemical Cytology* **8**, 617–638.

Silva, P. P. da & Branton D. (1970). Membrane splitting in freeze-etching. *Journal of Cell Biology* **45**, 598–605.

Sjöstrand, F. S. & Baker, R. F. (1958). Fixation by freeze drying for electron microscopy of tissue cells. *Journal of Ultrastructure Research* **1**, 239–246.

Steere, R. L. (1957). Electron microscopy of structural detail in frozen biological specimens. *Journal of Biophysical and Biochemical Cytology* **3**, 45–60.

Stowell, R. E. (Ed. (1965). *Cryobiology*, Federation Proceedings Supplement 15, Washington: Federation of American Societies for Experimental Biology.

Replicas and shadowing

Bradley, D. E. (1959). High resolution shadow-casting technique for the electron microscope using the simultaneous evaporation of platinum and carbon. *British Journal of Applied Physics* **10**, 198–203.

Bradley, D. E. (1965). In *Techniques for Electron Microscopy*, 2nd edn, Chap. 5 (Edited by D. H. Kay), Oxford: Blackwell Scientific Publications.

Pease, D. C. (1964). *Histological Techniques for Electron Microscopy*, 2nd edn, Chap. 10, New York and London: Academic Press.

Scanning electron microscopy

Boyde, A. & Wood, C. (1969). Preparation of animal tissues for surface-scanning electron microscopy. *Journal of Microscopy* **90**, 221–249.

Heywood, V. H. (1969). Scanning electron microscopy in the study of plant materials. *Micron* **1**, 1–14.

Oatley, C. W. (1966). The scanning electron microscope. *Scientific Progress, London* **54**, 483–495.

Oatley, C. W., Nixon, W. C. & Pease, R. F. W. (1965). Scanning electron microscopy. *Advances in Electronics and Electron Physics* **21**, 181–247.

High voltage electron microscopy

Cosslett, V. E. (1969). High voltage electron microscopy. *Quarterly Review of Biophysics* **2**, 95–133.

Dupouy, G. (1968). In *Advances in Optical and Electron Microscopy*, Vol. 2, pp. 167–250 (Edited by R. Barer and V. E. Cosslett), New York and London: Academic Press.

Dupouy, G., Perrier, F. & Durrieu, L. (1960). L'observation de la matière vivante au moyen d'un microscope électronique sous trè haute tension. *Compte rendu hebdomadaire des Séances de l'Académie des Sciences, Paris* **251**, 2836–2841.

*Hama, K. (1970). A stereoscope observation of tracheal epithelium of mouse by means of the high voltage electron microscope. *Journal of Cell Biology* **45**, 654–649.

*Hama, K. & Porter, K. R. (1969). An application of high voltage electron microscopy to the study of biological materials. *Journal de Microscopie* **8**, 149–158.

*Nagata, F., Hama, K. & Porter, K. R. (1969). Three-dimensional observation of biological specimens with high voltage electron microscope. *Journal of Electron Microscopy* **18**, 106–109.

Parsons, D. F. (1970). In *Some Biological Techniques in Electron Microscopy*, pp. 19–28 (Edited by D. F. Parsons), New York and London: Academic Press.

*Porter, K. R. & Hama, K. (1968). High voltage electron microscope study of tissue sections. *Journal of Cell Biology* **39**, 157 (Abs).

*Ris, H. (1969). Use of the high voltage electron microscope for the study of thick biological specimens. *Journal de Microscopie* **8**, 761–766.

Stereophotography and tilting specimen stages

Gray, E. G. & Willis, R. A. (1968). Problems of electron stereoscopy of biological tissue. *Journal of Cell Science* **3**, 309–326.

Wachtel, A. W., Gettner, M. E. and Ornstein, L. (1966). In *Physical Techniques in Biological Research*, Vol. III, Part A. p. 239 (Edited by A. W. Pollister), New York and London: Academic Press.

See also references marked with an asterisk under *High Voltage Electron Microscopy*

Photographic reinforcement technique

Friedman, M. H. (1970). A re-evaluation of the Markham rotation technique using model systems. *Journal of Ultrastructure Research* **32**, 226–236.

Markham, R. (1968). In *Methods in Virology*, Vol. 4, pp. 503–529 (Edited by Maramorosch and H. Koprowski), New York: Academic Press.

Markham, R., Frey, S. & Hills, G. H. (1963). Methods for the enhancement of image and detail and accentuation of structure in electron microscopy. *Virology* **20**, 88–102.

II

Routine Maintenance of the Electron Microscope

The number of electron microscopes throughout the world is increasing astronomically. These instruments have found uses not only in the biological sciences but also in metallurgy, geology, engineering, materials science, textiles, plastics and numerous other industries. They are used for research in basic physics, chemistry and biochemistry. Since the Apollo moon landings, they have been used in the investigation of lunar samples.

On the biological side, electron microscopes have become standard equipment in medical schools, departments of biological sciences, and biological research institutes. Agriculture and horticulture, marine biology and natural history have all benefited from ultrastructural studies.

The British Joint Committee for Electron Microscopy listed 766 transmission electron microscopes in Great Britain in early 1970. Of these, by far the largest number are of British manufacture (Associated Electrical Industries—known as 'AEI'—Harlow, Essex). Other manufacturers are Philips (The Netherlands), Siemens (West Germany), Jeolco (Japan), Hitachi (Japan), and Zeiss (West Germany).

The view ahead is open-ended. The era of electron microscopy in many fields is just beginning: pathology is a pertinent example. A few hospitals have already installed electron microscopes for use in the diagnosis of disease. Others will follow suit as more techniques for ultrastructural diagnosis are devised and as small, less expensive, medium resolution electron microscopes become available.

With this rapid proliferation of instruments the demand for trained personnel to run them has increased apace. Courses involving training in electron microscopic techniques are creeping into the curricula of universities, medical schools and technical colleges. The demand for experienced technicians to cope with the routine maintenance of electron microscopes has always far outstripped the supply, and this gap continues to widen. Because of this it is often necessary to employ completely inexperienced personnel and train them on the job.

It frequently occurs that a microscope is purchased when one or more workers in an institution have taken their research to a point where study with an electron microscope is the next logical step. The persons involved may never have had any practical experience with an electron microscope, and will be learning the techniques from the ground up. A technician may be employed shortly before or at the time of installation of the microscope to handle the routine maintenance and preparation of specimens.

Persons who will be in charge of a new instrument are often sent on a short course (of about 2 to 5 days duration) given by the manufacturers of the microscope. This gives the basic essentials of electron microscopy and routine care of the instrument. How much actual benefit is derived from such a course varies widely with the quality and duration of the course and the past experience of the person attending. Someone with a solid background in physics and electronics coupled with some practical experience with delicate instruments will of course benefit more from the instruction than someone with a poor command of such background subjects and little or no practical experience. For the latter, it will be very difficult to absorb all the pertinent details in a usable form within the duration of the course, particularly if there is no one at his place of employment to give him further guidance when he returns.

The present chapter is written with such a person in mind. It presupposes no specialized former training beyond courses in general biology and general chemistry. It does presuppose that the instrument in question is being maintained on a service contract and that only the daily routine maintenance is to be attended to by the technician in charge. At the end of the chapter references are given to books which deal with the maintenance of the microscope and 'troubleshooting' in more detail.

SERVICE CONTRACTS

The manufacturers of electron microscopes usually offer service contracts for the convenience of their customers. These are particularly useful when no one responsible for the microscope has had training in electronics and engineering. An electron microscope is a complicated, delicate and very expensive piece of engineering; it can be easily put out of kilter by inexperienced hands.

The cost of a service contract depends upon the services contracted for. The simplest contract entails only annual cleaning and checking of the instrument and costs in the region of £50 to £100 per year. A full contract providing for unlimited visits of the service engineer may cost as much as £500 per year. Between these extremes, arrangements can be made according to the expected needs of the user. Without such a contract the cost of a service engineer is in the region of £28 per day at present.

Obviously, whether to have a service contract and which one to choose depends upon how many times per year something is likely to go wrong with the microscope which the person in charge cannot put right. By the end of the warranty period (usually one year after installation) it should be relatively easy to judge whether or not a service contract is advisable. Servicing an electron microscope is a specialist job. This is particularly true for the more modern microscopes. In general, as microscope operation becomes simpler, the more complicated becomes its servicing. Most new users, particularly in the biological sciences, are quite happy to take out a service contract at expiry of the warranty and are wise not to discontinue it.

ALIGNMENT

It should be stated at once that the best source of information on the alignment and routine care of a particular electron microscope is the instruction manual provided by the manufacturer. Especially in those instruments designed solely for biological purposes, such as the AEI 801 electron microscope, certain of the alignment adjustments are carried out at the factory or when the instrument is installed and need no attention on the part of the operator. Types of mechanical controls differ from one microscope to another, and the sequence of adjustments may vary considerably. There are several ways of carrying out most of the procedures involved. The methods recommended by the manufacturer represent those particular routines which have proved most consistently valid with their instruments over a period of years. Any departure from these routines should be made only on the advice of the manufacturer's service representative.

Unfortunately, not all instruction manuals are outstanding for their clarity. This may be particularly true if the manual has been translated from another language. Also, the alignment routines given may be simply lists of manipulations to be performed, no attempt being made to inform the operator what the purpose of the individual manipulation is. Because of this the following discussion on alignment is included in order to give the reader an idea what this rather complicated process is all about. The description is deliberately oversimplified, on the assumption that one is better able to cope with complicated details later if one initially forms an overall picture into which to fit them. Once this has been done, the excellent discussions on the theory of alignment of the electron microscope by Siegel (1964), Sjöstrand (1967) and Meek (1970) should be studied.

The purpose of alignment is to get the parts within the electron microscope column positioned in relation to one another in such a way that the ultimate resolution of which the machine is capable can be attained. Misalignment results in image movements which cause blurring and reduce resolution. Ideally, the principal ray* of the electron beam should pass straight down the column, its axis of propagation being coincident with the mechanical and optical axes of the microscope.

There are three vertical axes which must be taken into consideration in aligning an electron microscope:

1. The mechanical axis, which passes through the filament tip, anode aperture centre, the aperture centres of the condenser, objective and projector lenses, and the viewing screen centre.

2. The magnetic axis (current centre) refers to that point on the image as seen on the fluorescent screen which remains stationary when the current through the coil of any imaging lens (i.e. objective and projector lenses) is varied at constant voltage. Points away from this centre will be seen to rotate about the centre as the current is varied. This centre represents the point on the image where fluctuations in lens current cause the least displacement of the image. The magnetic axes of the imaging lenses should correspond with one another and define together one optical axis of the instrument.

3. The voltage axis (voltage centre). This refers to the point on the image which remains motionless when the voltage applied to the gun is varied. Points lying peripheral to this will move radially with respect to the voltage centre when the voltage is increased or decreased. The voltage centre represents that point at which fluctuations in the high voltage supply create minimum displacement of the image. It is the second optical axis of the instrument.

Ideally, in order to attain the ultimate resolution of which the instrument is capable, the three axes of the electron microscope should correspond. In practice, however, the voltage and current centres seldom correspond exactly due to inaccuracies in machin-

* Principal ray: The central ray, which defines the direction of propagation of the radiation (Siegel, 1964).

ing the components. It is the manufacturer's business to see that the lack of correspondence is not so great as to impair the resolution quoted for the machine. Since the two optical axes are not in exact correspondence, it is necessary to choose which one of them shall be aligned with the mechanical axis. The alignment routine of most manufacturers entails aligning the current centre with the mechanical axis. (For high resolution work, however, some operators choose to align their microscopes with respect to the voltage centre.)

Alignment of most instruments usually proceeds from the top of the column downward. The fixed centre of the viewing screen is used as a reference point during the adjustments. First the electron gun (cathode plus anode) is aligned relative to the two condenser lenses (these lenses are usually fixed with respect to one another at the factory). The gun-condenser system (i.e. the illuminating system) is then aligned relative to the objective lens. Finally, the projector lenses (which are sometimes referred to as the 'intermediate' and projector, or as P1 and P2) are aligned relative to the objective. These various alignments are done by observing the movements of the image on the fluorescent screen as the current through the lenses is varied. The appropriate controls for lateral (translational) movement or axial tilting are then manipulated until image movement reaches a minimum at the fixed centre of the viewing screen.

The most critical part of the alignment procedure is getting the electron gun and condenser lenses lined up with the magnetic axis of the objective lens. If this is not accomplished within a very limited tolerance, spherical and chromatic aberration will be excessive. The alignment between the objective and the projector lenses is somewhat less critical. However, proper alignment between the two will ensure that the image will stay centered as the magnification is changed.

It should be determined either from the manufacturer's instruction manual or from his service representative:

1. Which parts of the complete alignment procedure have been done at the factory or at time of installation and should thereafter remain unaltered. (This may include alignment of two or more of the lenses with respect to one another. These are then referred to as fixed lenses.)

2. Which parts of the alignment procedure should be done only by the service engineer at his regular visits and/or when the column is dismantled for cleaning or major repair.

3. Which parts of the alignment procedure are necessary during day-to-day operation of the instrument.

It is the latter category which will be of main concern to the person in charge of routine maintenance. In modern biological instruments the daily procedures frequently include only the aligning of the illuminating system (i.e. aligning the gun with respect to the condenser lenses) and the centring of apertures. The specific details will, of course, vary from one machine to another.

X-RAY HAZARD

X-rays are generated during the operation of the electron microscope when the accelerated electrons strike any obstacle in their path. The anode aperture, the second condenser aperture, the specimen holder, objective aperture and holder and viewing screen are all points of impact. In order to protect operators from this hazard the column is carefully shielded, and leaded glass is fitted into the observation ports. During normal operation of the electron microscope, therefore, the risk from X-irradiation in a properly shielded and aligned machine is virtually nil. During certain stages of the complete alignment procedure, however, there is danger of excessive X-ray exposure unless certain stipulations regarding beam current and accelerating voltage are adhered to. These should be enumerated in the instruction manual, and must be scrupulously observed.

Two main points should be remembered: X-ray emission will reach a dangerous level if the machine is operated with the condenser aperture removed, or with the first condenser lens switched off.

It is always advisable to have a newly installed microscope monitored by a Geiger counter and to have personnel wear film badges for the first hundred hours or so of operation to ensure that the shielding of the machine has been properly carried out at the factory (Meek, 1970).

TESTING THE PERFORMANCE OF THE MICROSCOPE ONCE ALIGNMENT HAS BEEN COMPLETED

This should be done routinely after aligning the column to make sure that (1) any errors in alignment which may have occurred are not sufficient to adversely affect the desired resolution, and (2) that serious contamination is not present. Testing can be done using a thin carbon film which contains holes. The width of the minimum diffraction fringe measured on a plate taken at the top magnification of the instrument will indicate the resolution of which the machine is capable under the prevailing operating conditions. If the fringe width indicates a resolution insufficient for the study to be made, the alignment procedure should be repeated. If the fringe is overfocused in one direction and underfocused at right angles to it, this indicates that objective astigmatism is present and should be corrected. If contamination in the specimen area is not serious it can be compensated for with the objective astigmatism corrector as described in Chapter 5. If the astigmatism is uncorrectable, the objective aperture should be changed. If still uncorrectable, the source of contamination in the specimen area (e.g. on the specimen holder, objective aperture rod, or screening tube) must be located and removed.

Slow movement of the specimen (drift) will cause blurring of the image but is not always visually obvious. Drift can be checked for by photographing a hole twice on the same plate at an interval of 2 min. If drift is occurring the hole will show lateral displacement. This test will also show the rate at which the specimen is being contaminated. The difference in radius of the hole in ångstrom units after a 2 min exposure to the beam, divided by the the time in seconds, will give the contamination rate in Å/s. The contamination rate in a microscope without an anticontamination device should not exceed 1 Å/s. An anticontaminator should reduce the rate to between 0·1 and 0·01 Å/s or less (Meek, 1970).

MAGNIFICATION CALIBRATION

The actual magnification of the microscope may differ from that on the control settings by a considerable amount, and at best only corresponds to the settings to within ± 5 per cent. This is due to a number of factors including variations in lens current and high voltage, and variations in the position of the specimen along the microscope axis. Both the accurate determination of microscope resolution, and precise measurements of structures in micrographs are dependent upon accurate calibration of the magnification.

The magnification of the instrument for each control setting should be re-calibrated from time to time, especially before high-resolution work is to be done, or if accurate measurements are to be made. Calibration is conveniently done by photographing a replica of a diffraction grating, the spacing of the lines of which is accurately known. Photographs of the replica are taken at each magnification setting, and a number of measurements made and averaged for each setting. Diffraction grating replicas may be made in the laboratory, but are difficult to prepare accurately. It is simpler to obtain one from suppliers of electron microscope materials. If good care is taken of it, a replica will last for some years.

Calibration with diffraction grating replicas becomes difficult at higher magnifications, and other types of calibration specimen may be preferred. The various methods of calibration are discussed in detail by Meek (1970) and Sjöstrand (1967).

SOME GENERAL INFORMATION ABOUT MAINTENANCE PROCEDURES

When to Leave Well Enough Alone

If an instrument is performing properly (i.e. giving the resolution required for the study involved) it is the best policy to *leave it alone* until there is some very good reason for doing something to it. Every time the column is opened there is the chance of dirt getting in. When the column has been opened, it may take several days to degas it sufficiently for the optimum vacuum to be attained. Every time the column is taken down ('stripped') there is the possibility of introducing severe misalignment or damaging delicate parts while getting it back together. Time between cleanings will vary according to the vacuum routinely employed, the cleanliness of laboratory air and water, and the number of photographic plates or film used. However, unless an instrument is receiving round-the-clock use, stripping and cleaning the entire column will probably not be necessary at shorter intervals than 12 months. Between complete strippings (which are done by the service engineer) routine cleaning can probably be confined to those parts of the column lying above the objective lens, such as the gun housing, the cathode shield, the anode, and those parts which can be removed from the column without opening it up, such as the condenser and objective apertures and the specimen holder. In other words, it should not be necessary to remove the condenser lenses from the column between major overhauls.

Any part which goes into the column must be scrupulously clean and thoroughly dry and should never be touched by the fingers. Clean specimen holders and aperture rods should be stored in the presence of a desiccant until placed in the microscope. Grease from the fingers and other contaminants will be vapourized under vacuum conditions and contribute to the contamination of the interior surfaces of the column. Compounds produced by bombardment of the contaminating material by the electron beam may be charged sufficiently to distort the image and reduce resolution.

Contamination on the portion of the specimen holder which is immersed in the objective lens field can be particularly bothersome in this respect. If photographic plates or roll film have not been properly dried they will release water vapour into the column. Plates should be dried in a vacuum desiccator over some such desiccant as phosphorous pentoxide for at least two hours before being placed in the column; roll film somewhat longer.

After removal from the desiccator the plates must be placed into the microscope column at once, since five minutes' exposure to normal atmospheric pressure will completely reverse the effects of desiccation. Some laboratories where high resolution work is done let dry nitrogen into the plate desiccator to break the vacuum rather than air. The nitrogen forms a layer over the plates and prevents the emulsion from picking up moisture during the trip from desiccator to microscope. The microscope column itself may also be returned to atmospheric pressure by the introduction of dry nitrogen for the same reason.

Clean white nylon or cotton gloves or lint-free paper should be used when handling parts. Parts temporarily removed from the column should be placed on lint-free paper in a dust-free place. When open, the top of the column should be covered, preferably with a clean metal plate, to prevent particles of dust falling into the column.

Ideally an electron microscope laboratory should be so designed that dust and other contamination are kept to a minimum. Walls should be of smooth surface texture to discourage accumulation of dust. A ventilation system should be installed which provides for the movement of filtered air.

Maintenance Schedules and Records

A maintenance schedule should be worked out to fit the needs of the particular laboratory. Certain things should be done daily, others weekly or monthly. If the microscope is in constant use, the objective apertures may have to be changed daily. If the contamination rate is high or if high resolution work is being performed, several changes of individual apertures may be necessary in one day. A clean objective aperture rod loaded with clean apertures should always be kept in readiness. It should be kept dry and free from dust in a clean, sealed tube.

Desiccants for the microscope rotary pump and the plate desiccator should be changed at least once a week. Other jobs, such as cleaning the gun housing and anode can be done when a blown filament necessitates opening the column. If an operator is in a hurry to resume work on the microscope, however, the cleaning can be done another time, at approximately monthly intervals.

Monitoring of certain currents and voltages should also be done monthly and a record kept for comparison when the service engineer does his routine checking. The contamination rate should be checked monthly to make sure it is not increasing. Dates of filament changes should be recorded. This will give some idea when a filament may blow. If filament life is very short it may indicate that the vacuum attainable with the diffusion pump is deteriorating. Details of all breakdowns which occur should also be recorded and kept for reference.

Changing the Filament

Filament life varies with operating conditions. A reasonable filament life is considered to be about 25 hours. Evaporation of metal from its surface weakens it, and oxidation may occur due to residual gas and water vapour molecules in the column.

In order to save time, a spare cathode assembly should be kept in readiness for insertion into the column when the filament burns out. It must be clean and the filament accurately inserted and centered according to directions in the instruction manual. The position of the filament tip with respect to the cathode shield aperture is critical and varies from one instrument to another.

When a filament burns out, the beam disappears and the reading on the filament current meter falls. There may be some flickering of the beam for a short while before this occurs, or it can happen without warning. The filament current knob should be turned down and the filament switch put to the off position. The high voltage should be turned off, THE DIFFUSION PUMP SHUT OFF FROM THE COLUMN and air admitted to the column as directed in the instruction manual. When the column has reached atmospheric pressure the gun housing may be opened.

WARNING: THERE MAY BE A RESIDUAL CHARGE ON THE GUN. To avoid a severe shock, remove the charge by touching simultaneously the interior of the column wall and the cathode shield with an insulated screwdriver.

WARNING: THE CATHODE ASSEMBLY MAY BE VERY HOT. It should be removed carefully, using a clean cloth or tissue, and set aside. The spare cathode assembly is now inserted into the gun chamber. Check that the gun chamber and porcelain insulator are clean and dust free, and that the anode aperture appears to be centred. If the anode and condenser apertures are dirty they should be cleaned and replaced respectively at this time. Then close the gun chamber and pump out the column. When the requisite vacuum has been attained, the filament may be switched on. Unless the filament has been wrongly centered or has moved when warming up, a beam should appear. If it does not, the column must be re-opened to check the position of the filament. Filaments are usually 'pre-flashed' at the factory to remove stresses so that movement will be less likely to occur on heating by the filament current. If the filament moves enough to touch the cathode shield it will rapidly burn out again.

If a beam is obtained and if the gun is correctly aligned, the illumination should rise steadily to a maximum brightness and remain at that intensity when the current is increased. The centreing of the condenser aperture should be checked and the condenser and objective astigmatism corrected. Then normal operation may be resumed.

The filament assembly which contains the blown filament should be cleaned and a new filament inserted and centred ready for use. The inside of the cathode shield will be covered by tungsten evaporated from the filament as well as the usual hydrocarbon contaminants. If the contamination is not too heavy it may be removed by sonication as described in the section on Cleaning Procedures.

Inspection of the blown filament will show whether normal evaporation, too high gas pressure in the column, or overheating caused the burn-out (Fig. 16). The burn-out will usually occur on

Fig. 16. Appearance of filament after burning out. (*a*) Cause: normal evaporation. (*b*) Cause: excessive gas in column. (*c*) Cause: overheating.

one side of the tip, whatever the cause. The edges of the gap will be smooth if the break is due to normal evaporation, jagged if it is due to gas attack, beaded if caused by overheating.

Centring the Condenser and Objective Apertures

If the condenser aperture is incorrectly centred the beam will sweep across the field instead of expanding and contracting evenly and concentrically about the centre of the viewing screen when the condenser fine focus knob is rotated. Centre the aperture with the appropriate controls. If the beam is elliptical instead of round after centring the aperture, reset the condenser astigmatism controls to compensate for this.

If the objective aperture drifts off centre by more than one third of its diameter it will invalidate the astigmatism correction for the objective lens, and may even cut off the beam. Its position should be checked occasionally.

Cleaning Procedures for the Electron Microscope

There are a number of different methods of cleaning the various components of the electron microscope column, so that detailed practice varies from one laboratory to another. Before attempting to clean any parts of the microscope for the first time, detailed advice should be sought from the manufacturer's service representative.

It is most important to learn what NOT to do. In particular, no components within the lenses should be touched by inexperienced hands. In all probability the manufacturer's representative will as a matter of course demonstrate the methods to be employed.

A common cleaning routine which gives satisfactory results involves sonication of those parts which are small enough to fit into the sonicator, and careful polishing of those which are too large to sonicate, or which are very heavily contaminated.

Polishing.

Polishing of heavily contaminated parts, or parts too large to sonicate, may be done with a high-grade non-abrasive silver polish, such as Silvo, on a soft cloth. Abrasives such as emery or carborundum should never be used. The cotton buds on applicator sticks used for infant hygiene make handy polishing implements for areas which are difficult to reach with a cloth. Polishing should be followed by thorough washing in detergent solution and scrupulous rinsing with distilled water. Drying may be effected with a soft cloth or in a drying oven. The drying temperature should not exceed 40°C.

Sonication

A sonicator is basically a small tank which is caused to vibrate at high frequency (10 to 100 kHz)* by transduction of an electric

* kHz = kilohertz = 1000 cycles per second.

current. The vibration caused in the walls of the tank is transmitted to a liquid (usually tap water) which is contained in the tank to a depth of approximately 1 inch. The parts to be cleaned are placed in a container and covered with a cleaning solution. The container is placed in the tank, its bottom in contact with the tap water. The container is suspended from a clamp stand in such a way that it does not contact the walls of the tank. Vibration set up in the water is transmitted through the glass walls of the container to the liquid surrounding the parts to be cleaned. This vibration gently but effectively removes contaminating material from the parts.

WARNING: *Ultrasonic vibration can injure human tissues, so do not touch the sonicator while it is in operation.*

It is important to use cleaning solvents which can be successfully removed during subsequent rinsing; otherwise residual solvent molecules will evaporate into the column and contribute to contamination.

WARNING: *Solvents such as acetone or ether should never be used in a sonicator.* The vibration steps up evaporation and increases the danger of fire, explosion and toxic effects.

A satisfactory and safe pair of cleaning solutions to be used in succession is recommended by the manufacturers of the AEI line of electron microscopes. These are: (1) a lipid solvent, 1, 1, 1-trichloroethane (Inhibisol); and (2) and ammoniated detergent which acts as a weak etching agent (Quadralene instrument cleaner).

The following routine has been found satisfactory:

1. Sonicate for 1 min in Inhibisol.
2. Blow dry with room temperature air from a filtered compressed air line.
3. Sonicate for 2 min in 10 per cent Quadralene in cold water.
4. Wash thoroughly with distilled water.
5. Blow dry.
6. Wash for one minute in clean Inhibisol.

De-ionized water should not be used in the cleaning procedure, and distilled water should be produced by a glass still rather than a metal one.

Cleaning of Lens Apertures

After preliminary sonication, as described above, the molybdenum apertures used in the condenser and objective lenses should be further cleaned by heating in a vacuum. This can be conveniently done in a vacuum coating unit by placing the apertures dished side down on a molybdenum 'boat' which is suspended between the low-voltage high-current terminals. (The 'boat' should first be cleaned by heating (flashing) as described below before the apertures are put in.) Wooden cocktail sticks or toothpicks make handy tools for gently removing aperture discs from their carriers if they will not simply shake out.

The apertures are heated to yellow heat (approximately 1200°C) by passing a current of about 65 A through the molybdenum boat for about 5 seconds. The apertures are then allowed to cool to room temperature, removed from the coating unit, and checked under a microscope with the 10× objective to assure that they are clean. If contamination is still present, the heating process can be repeated. If the contamination is still present after three heatings the apertures should be discarded.

Care of the Pumps

The oil in the diffusion pump will need changing occasionally. This is usually done by the service engineer during major overhauls. If a failure of the cooling system to the diffusion pump occurs, overheating may result in 'cracking' of the oil, which will then have to be replaced. ('Cracking' means that the long-chain hydrocarbon molecules which have a high boiling point are broken down into smaller molecules of lower boiling point.)

The oil in the rotary pump should be kept at the level indicated on the sight glass, and changed when it becomes cloudy due to contamination with water molecules.

Desiccants

The desiccants used in plate desiccators and rotary pumps will have to be renewed at frequent intervals. The appearance of the material will frequently tell you when a change is necessary, i.e. indicator crystals in silica gel will turn from blue to pink; phosphorous pentoxide will cake together with moisture.

WARNING: If phosphorus pentoxide is used, great care must be taken not to expose the fresh material to water as *it will explode*. Material which has already absorbed a large amount of water vapour and become caked together may be washed out of its container under a tap, but even this should be done very cautiously.

Water Filters

If the microscope is fitted with a device for filtering the water for the cooling system this will need periodic cleaning to ensure adequate water flow. Blockage of the water supply to a lens quickly causes overheating and damage to the coils.

Fan Filters

If the power unit has a fan fitted with a filter this too will need cleaning or replacement at intervals. This is very important, since an uninterrupted supply of cool air is essential.

Renewing the Phosphor on the Fluorescent Screen

Every thousand operating hours or so, the fluorescent material coating the viewing screen will need renewal. The microscope manufacturers will probably recommend a suitable phosphor and provide instructions for its use. If they do not, see Meek (1970). Very fine-grain screens are coming into increasing use. These are difficult to make properly and are best obtained through the microscope manufacturers.

Warning: Since the zinc and cadmium sulphides usually used in the phosphors are poisonous, the screen and the coating material should be handled with great care to avoid dispersal of the phosphor in the air. The hands must be carefully washed after the recoating process is finished.

Power Supply

Certain voltages and currents in the power unit must be monitored from time to time with the appropriate equipment, and a record kept of their readings in order to facilitate the diagnosis of trouble if anything goes wrong. The techniques for performing these tests should be learned under the supervision of the service representative. A power supply providing 100,000 V CAN BE LETHAL TO THE INEXPERIENCED. The latter should be allowed only to read existing meters.

Reference Books

Anyone in charge of an electron microscope laboratory should have at his fingertips certain books which discuss in depth routine maintenance and specimen preparative procedures. A list of such books is given at the end of this chapter.

Some Common Problems and What to do About Them

Table 6 lists some of the more common symptoms which trouble the microscopist. The list is by no means exhaustive and only the more likely causes are mentioned.

Table 6. Some common problems and what to do about them

Symptom	Possible causes	Course of action
No beam when filament is energized	1. High voltage off	Switch on high voltage
	2. Filament blown	Check beam current metre; if no current change filament
	3. Fluorescent screen up	Lower screen
	4. Specimen chamber closed off from column	Open specimen chamber to column
	5. Something else obstructing beam	Move specimen to ensure grid bar not obstructing beam. Are two grids superimposed in holder? Check position of condenser and objective apertures
	6. Condenser 2 defocused	Focus condenser 2
	7. Magnification not at lowest setting	Return to lowest magnification
	8. Specimen traverse rods at end of runs	Recentre traverse rods
	9. Column misaligned	Realign column
Beam disappears during operation	1. High voltage cut off due to low water pressure or vacuum leak	Determine cause and correct
	2. Filament blown	Change filament
	3. Specimen traverse rods at end of run	Recentre rods
	4. Some alignment control inadvertently moved	Check control settings
	5. Objective aperture has shifted	Recentre aperture
Beam cross section is elliptical or distorted	1. Condenser astigmatism needs correction	Correct astigmatism
	2. Contamination of condenser too great for astigmatism corrector to compensate	Check condenser aperture, aperture rod and screening tube for contamination

Symptom	Possible causes	Course of action
Flickering of beam	1. Filament about to blow	Wait a minute
	2. Contamination associated with gun or gun chamber	Clean gun area
	3. Electrical instability	Call service engineer
Reduced illumination	1. Gun misaligned	Adjust anode or gun traverse
	2. Filament too far from cathode shield aperture	Adjust filament height
Beam sweeps across screen when condenser current is varied	Condenser aperture off centre	Recentre condenser aperture
Image moves across screen when objective focus is varied	Condenser lenses not properly aligned with respect to objective lens	Correct alignment
Image moves across screen when magnification is varied	Projector lenses not properly aligned with respect to rest of column	Correct alignment
Image will not focus	1. Coarse focus knob out of focus	Focus progressively from coarse to fine knobs
	2. Astigmatism of objective lens	Correct astigmatism or insert clean apertures and correct astigmatism
	3. Specimen holder not properly seated	Check position of specimen holder
	4. Electrical fault	Call service engineer
Blurring of diffraction fringe in one direction around a hole in a carbon film	Mechanical vibration near microscope or a unidirectional varying magnetic field	Determine source of vibration or field and remove
Blurring of diffraction fringe all around hole in carbon film	Instability in lens current or accelerating voltage	Call service engineer

Symptom	Possible causes	Course of action
Image drift	1. Thermal expansion of specimen or grid in beam	Scan area with low beam. It may settle down
	2. Torn specimen	Try another area on grid or another specimen
	3. Dirt on specimen	Discard specimen
	4. Dirty specimen holders	Clean holders
	5. Defective specimen stage	Call service engineer
Short filament life	1. Beam current too high	Do not increase current past saturation
	2. Insufficient vacuum	Check for leaks by manipulating various controls and watching vacuum gauge. Also, pumps may be dirty; engineer may be required
	3. Filament position incorrect with respect to cathode shield aperture	Adjust position of filament
Discharges in gun	1. Insufficient vacuum	Check for leaks etc. as above
	2. Contamination in gun chamber	Remove contamination
High contamination rate	1. Dirty specimen holders	Clean specimen holders
	2. Insufficient vacuum	Check for leaks etc. as above
	3. 'Cracked' oil	Clean pumps, instill fresh oil (service engineer)
Layer of contamination visible on fluorescent screen and/or viewing port	Oil in column due to failure of valve between diffusion pump and column; mains electricity failure; or failure of pump cooling system	Call service engineer to strip and clean column and diffusion pump

REFERENCES

General

Agar, A. W. (1965). In *Techniques for Electron Microscopy* (Edited by D. H. Kay), Oxford: Blackwell Scientific Publications.

Siegel, B. M. (1964). In *Modern Developments in Electron Microscopy*, pp. 66–70 (Edited by B. M. Siegel), New York and London: Academic Press.

Recommended reference books for use in an electron microscope laboratory

Kay, D. H. (Ed.), (1965). *Techniques for Electron Microscopy*, Oxford: Blackwells Scientific Publications.

Meek, G. A. (1970). *Practical Electron Microscopy for Biologists*. See particularly pp. 242–243; 268–280; 335–339, 372, London and New York: Wiley Interscience.

Pease, D. C. (1964). *Histological Techniques for Electron Microscopy*, 2nd edn, New York and London: Academic Press.

Sjöstrand, F. S. (1967). *Electron Microscopy of Cells and Tissues*, Vol. 1, pp. 311–325, 362–366, New York and London: Academic Press.

Appendix

I. HOW TO ALLOW FOR DIFFERENT HYDRATION STATES OF CHEMICALS

The amount of a given chemical in a recipe may have to be altered according to the state of hydration of the reagent available in the laboratory. For example if the recipe calls for 2·0 g of anhydrous sodium acetate ($NaC_2H_3O_2$, molecular weight 82·04) and the reagent on the shelf is hydrated ($NaC_2H_3O_2 3H_2O$, molecular weight 136·09) one will obviously have to use $136/84 \times 2{\cdot}0$ g of the hydrate, or 3·2 g.

Unfortunately, authors do not always specify the hydration state of the chemical. If hydration state is not specified, one is forced to assume either the anhydrous state (if the chemical is commercially available in that form) or else the lowest degree of hydration which is commercially available.

The molecular weights, possible states of hydration, and other physical constants of chemicals can be found in the *Handbook of Chemistry and Physics* published by the Chemical Rubber Publishing Company, Cleveland, Ohio. It is affectionately known as the 'Rubber Bible' and is a 'must' in every laboratory.

II. MOLARITIES OF ACID AND ALKALI USUALLY USED FOR ADJUSTING pH IN THE SUCCEEDING FIXATIVE AND BUFFER RECIPIES

The following calculations do not result in exact molarities, which can only be determined by titration methods. However, they are sufficiently accurate for most electron microscopical procedures.

Sodium hydroxide:

M. W.=40, therefore 4 per cent w/v =1 M
1 ml of 1 M diluted to 10 ml with distilled water=0·1 M
1 ml of 0·1 M diluted to 10 ml =0·01 M

The resulting solutions should be stored in plastic bottles, since alkali will etch glass and silicates will go into solution.

Concentrated hydrochloric acid may come in either of two molarities, 11·65 M being the most common. In diluting the acid, always work at a sink and carefully add acid to water, not vice versa.

Hydrochloric acid:

Sp. gr. 1·18=36 per cent w/w	=11·65 M
0·86 ml diluted to 10 ml with distilled water	=1 M
1 ml of 1 M diluted to 10 ml	=0·1 M
1 ml of 0·1 M diluted to 10 ml	=0·01 M

Hydrochloric acid:

Sp. gr. 1·16=32 per cent w/w	=10·2 M
0·98 ml diluted to 10 ml	=1 M
1 ml of 1 M diluted to 10 ml	=0·1 M
1 ml of 0·1 M diluted to 10 ml	=0·01 M

III. FIXATIVES

All stock solutions should be kept in a refrigerator (4°C) unless otherwise specified. Some solutions, particularly phosphate buffers and sucrose or glucose soon develop bacterial or fungal growth. Check for this by rotating the bottle and watching for a swirl of contaminating material at the bottom.

The fixatives in this section are generally buffered to a pH of 7·2 to 7·4. For highly hydrated material such as embryonic tissue, protozoans and invertebrates a pH of 8·0 to 8·5 has been said to give a better preservation.

Osmium Fixatives

One hour is the fixation time generally recommended for osmium fixatives.

Osmium fixatives should be made up a day before use, since osmium tetroxide dissolves very slowly. The freshly made fixative may be left in the dark at room temperature overnight, then placed in an ice bath in the refrigerator next morning.

Osmium tetroxide should be used only in a fume cupboard. Since the addition of osmium to a solution does not alter the pH, the pH of the solution may be adjusted outside the fume cupboard before adding the osmium. Osmium solutions are photosensitive and should be kept in brown glass bottles.

Since osmium tetroxide is easily reduced, the use of clean glassware is essential. Even if the fixative is kept in the refrigerator, reduction is likely to occur spontaneously within a few days, because of reactions between osmium tetroxide and other components in the fixative. (Osmium tetroxide dissolved in distilled water keeps much longer.) If the fixative has turned brown or reddish it should be discarded. Some osmium fixatives may be kept indefinitely if frozen. Others may be found to contain a precipitate and be of questionable value after freezing.

1. *Millonig's fixative* (Millonig, 1962).

Solution A:	2·26 per cent $NaH_2PO_4.H_2O$	Stable for a week or more.
Solution B:	2·52 per cent NaOH	Store in polythene bottle, stable indefinitely at room temperature.
Solution C:	5·4 per cent glucose	Will soon become contaminated even at 4°C.
Solution D:	41·5 ml of Solution A 8·5 ml of Solution B	
Fixative:	45 ml of Solution D 5 ml of Solution C 0·5 g OsO_4	

This very popular fixative is routinely used in many laboratories. It is said to preserve more protein and glycogen than the other osmium fixatives, and to give more uniform fixation throughout the block of tissue. The fixative will keep for some weeks if frozen. *Do not fill the fixative bottle too full if it is to be frozen.* The fluid will expand on freezing and may break the bottle. To thaw, warm the fixative gradually by running cool tap water over the bottle at first, then gradually warming up the water. The fixative is ready for use as soon as the last trace of ice has vanished from the fixative.

2. *Palade's fixative* (Palade, 1952)

Stock buffer solution: (stable for months)	14·7 g sodium veronal (barbital) 9·7 g sodium acetate Distilled water to make 500 ml
Stock osmium solution:	2 per cent osmium tetroxide in distilled water. (This aqueous solution will keep for some weeks if refrigerated, unlike the fixative).
Fixative:	5 ml stock buffer solution 5 ml 0·1 M HCl 2·5 ml distilled water 12·5 ml 2 per cent osmium tetroxide

This is the original osmium fixative developed for electron microscopy by Palade. It is still used in many laboratories throughout the world.

3. *Caulfield's fixative* (Caulfield, 1957)

This is simply Palade's fixative with 4·5 per cent sucrose added to increase the osmolarity of the fixative.

4. *Osmium fixative in s-collidine buffer* (Bennett and Luft, 1959)

Stock buffer solution: 2·67 ml *s*-collodine
50 ml distilled water
9·0 ml 1·0 M HCl
Dilute to 100 ml with distilled water

This yields a particularly powerful 0·2 M buffer solution with a pH of 7·4 to 7·5. It is stable indefinitely and can be used with osmium, permanganate and aldehyde fixatives. It tends to reduce osmium rather rapidly, however, and the fixative should be used within 3 to 4 days of preparation.

Fixative: One part buffer
Two parts 2 per cent aqueous osmium tetroxide

This yields 1·33 per cent osmium tetroxide in 0 067 M buffer.

This fixative was not widely used at first since commercially available *s*-collidine contained impurities which had to be removed by rather laborious methods. Most suppliers of materials for electron microscopy now offer a pure-grade *s*-collidine specially for the purpose.

Aldehyde Fixatives

Two to four hours is the fixation time generally recommended. Four hours is preferable for glutaraldehyde. Veronal acetate buffers cannot be used with aldehyde fixatives, as a Canizarro-type reaction occurs between aldehyde and buffer which destroys the buffering capacity.

Glutaraldehyde Fixatives

The purity and pH of the glutaraldehyde stock solution is important. Glutaraldehyde is subject to air oxidation and polymerization when in concentrations of 25 per cent and above. Products of these reactions accumulate rapidly unless the solution has been stabilized, and are detrimental to fixation. The pH of the solution drops because glutaric acid is formed. Sabatini *et al.* (1964) have found that if the pH of the stock solution falls below 3·5 fixation is poor. Suppliers of 'E.M. grade stabilized' glutaraldehyde are listed on page 220.

1. *Glutaraldehyde in Cacodylate Buffer*

Stock buffer solution of 0·2 M sodium cacodylate:
10·7 g sodium cacodylate, diluted to 250 ml with distilled water. Store at 4°C.

Stock glutaraldehyde solution:
25 per cent (or 50 per cent) solution of specially stabilized E.M. grade glutaraldehyde (provided as a 25 per cent or 50 per cent solution by the manufacturers). Store at 4° C.

Fixative: 8 ml of 25 per cent glutaraldehyde (or 4 ml of 50 per cent)
25 ml of 0·2 M sodium cacodylate
Adjust to pH 7·3 to 7·4 with 1 M HCl. Dilute to 50 ml with distilled water.

This yields 4 per cent glutaraldehyde in 0·1 M buffer, which is hypertonic. This percentage was recommended by Sabatini *et al.* (1963) who introduced glutaraldehyde as a fixative for electron microscopy. Many workers now prefer a lower concentration, usually in the neighbourhood of 2 per cent. This can be achieved by using 4 ml of 25 per cent glutaraldehyde in the above recipe.

2. *Glutaraldehyde Fixative in Millonig's Phosphate Buffer*

Stock solutions A: 2·26 per cent $NaH_2PO_4.H_2O$
B: 2·52 per cent sodium hydroxide
C: 25 per cent stock solution of glutaraldehyde

Fixative:
8 ml of 25 per cent glutaraldehyde (or 4 ml of 50 per cent)
32 ml of solution A
Adjust pH to 7·3 to 7·4 with solution B. Dilute to 50 ml with distilled water. (No glucose is added to this fixative since it is already hypertonic).

3. *Glutaraldehyde in s-collidine Buffer* (Weibel, 1969)

In order to reduce the tonicity of the fixative to roughly that of blood plasma Weibel (1969) recommends 1·5 per cent glutaraldehyde in *s*-collidine buffer. The solution is made up as follows:
27 ml 0·2 M *s*-collidine solution
3 ml 25 per cent glutaraldehyde
Check the pH, adjust if necessary to 7·2 to 7·4 and dilute to 50 ml

Paraformaldehyde fixatives

It is essential to use Analar paraformaldehyde powder when preparing formaldehyde fixatives for use in electron microscopy. Commercial 37 per cent solutions contain methanol as a preservative. This is detrimental to fixation. The fixative should be made fresh each time so that oxidation does not cause formation of formic acid.

1. *Paraformaldehyde in Millonig's buffer*

(This is a modification of Pease's recipe 1964).
Stock solutions A: 2·26 per cent $NaH_2PO_4.H_2O$
B: 2·52 per cent sodium hydroxide

Fixative: 41·5 ml solution A
8·5 ml solution B
Heat to 60° to 80° (covered to avoid evaporation). Add 2·0 g Analar paraformaldehyde powder. Adjust pH to 7·2 to 7·4. Stir until dissolved. Filter. Cool.

This yields a 4 per cent formaldehyde fixative. If a less hypertonic mixture is desired reduce the amount of paraformaldehyde powder accordingly.

2. *Paraformaldehyde-glutaraldehyde fixative* (Karnovsky, 1965)

1. Dissolve 2 g paraformaldehyde powder in 25 ml distilled water by heating to 60° to 70° and stirring
2. Add 1 to 3 drops 1 M sodium hydroxide with stirring until solution clears (slight milkiness may persist)
3. Allow to cool
4. Add 5 ml of 50 per cent glutaraldehyde
5. Increase volume to 50 ml with 0·2 M cacodylate *or* phosphate buffer, pH 7·4 to 7·6. The final pH will be 7·2. If cacodylate is used, add 25 mg $CaCl_2$ (anhydrous)

Fix for 2½ to 5 hours, wash 3 to 12 hours in 0·1 M buffer, post-fix in osmium tetroxide.

Although this fixative is highly hypertonic, its originator did not find shrinkage to be a problem.

Permanganate Fixatives

Since permanganate fixatives have largely fallen out of favour for electron microscopy, only Luft's classical recipe is given.

Solution A: 1·2 per cent potassium permanganate. Store in a well-filled stoppered brown glass bottle at 4°C (preferably prepare fresh solutions as needed)

Solution B: Palade's buffer:
14·7 g sodium veronal (barbital)
9·7 g sodium acetate
Distilled water to make 500 ml

Fixative: 12·5 ml 1·2 per cent $KMnO_4$
5·0 ml Palade's buffer
2·5 ml distilled water
5·0 ml 0·1 N HCl
Adjust pH to 7·3 to 7·4 with HCl

This gives a final concentration of potassium permanganate of 0·6 per cent. Higher concentrations tend to be more destructive to the tissues. Fixation time should not exceed 1 hour.

IV. RINSING AND HOLDING BUFFERS

These serve three purposes:

1. Removal of excess fixative from the tissue before dehydration begins. A quick rinse in distilled water would probably serve as well, and there is disagreement about the advisability of rinsing at all.
2. Removal of excess primary fixative before post-fixation in osmium tetroxide. There is disagreement about the necessity of this step and many workers omit it altogether—(see discussion in Chapter 2, p. 32).
3. Provision of a medium in which to store aldehyde-fixed tissue for an indefinite time until dehydration or post-osmication can conveniently be performed. This is permissible for certain routine work, but progressive leaching of certain materials from the tissue may occur. Some workers prefer to store the tissue in the primary fixative, but this is controversial.

The rinsing and holding buffers are made up from the stock buffer solutions used for preparing the fixative. Their pH should be 7·2 to 7·4. Their osmolarity is largely a matter of personal preference in the present state of knowledge (see discussion on osmolarity in Chapter 2). Sucrose or glucose may be used to adjust the osmolarity. If exact osmolarities are desired a good osmometer is essential. However, the total particle concentration may be used as a rough indication of osmolarity.

For example, to obtain 100 ml of roughly 0·3 osmolar solution of sodium cacodylate buffer: Adjust the pH of 50 ml of 0·2 M cacodylate to pH 7·2 to 7·4 with 0·1 M HCl, noting the exact amount of 0·1 M HCl necessary to do this. Calculate the particle concentration that both the sodium cacodylate and the HCl will contribute when the solution is diluted to 100 ml. This will be less than 0·3, so increase the particle concentration by adding the appropriate amount of 2 M sucrose before dilution.

In practice, the approximate amounts of the ingredients required are:

Ingredient	Molarity	Number of particles per molecule	Amount needed	Particle concentration after dilution to 100 ml
Sodium cacodylate	0·2	2	50 ml	0·2 M
HCl	0·1	2	8	0·016
Sucrose	2·0	1	4·2	0·084
Water			37·8	
			100·0 ml	0·3M

The osmolarity of such a mixture determined by the freezing point depression method is in the region of 0·29 osmoles.

V. EMBEDDING MEDIA

Only the most popular recipes for embedding media are given below. For other media and further details consult Pease (1964, Chapter 4), Glauert (1965) and references at the end of Chapters 2 and 8. Since the same ingredients may have several names, Table 7 has been included for the sake of clarity.

Table 7. Equivalent designations for embedding chemicals

Name	Equivalent	Function
Araldite CY 212	Araldite 502 (nearest U.S. equivalent)	Embedding resin
Dodecenyl succinic anhydride	HY964, formerly 964 B (U.K.); DDSA (U.S.)	Hardener
Tridimethylaminomethyl phenol*	DY064, formerly 964 C (U.K.); DMP 30 (U.S.)	Catalyst
Benzyl dimethylamine*	DY062 (U.K.); BDMA (U.S)	Catalyst
2, 4-dichlorobenzyol peroxide compounded with dibutyl phthalate	Luperco CDB (U.S.)	Catalyst
Methyl nadic anhydride	MNA; NMA; methyl-endo methylene phathalic anhydride	Hardener
1, 2-epoxypropane	Propylene oxide (U.S.)	Facilitates penetration of embedding medium

* May be used interchangeably.

Araldite Embedding

After dehydration in ethanol the tissue is passed through epoxypropane into the embedding mixture and left from 2 hours to overnight at room temperature. Overnight is probably safer from the point of view of ensuring complete infiltration. Convenient trays for infiltration can be made from plastic ice cube trays. After infiltration the tissue is placed in fresh embedding medium which is polymerized in a 60°C oven for about 24 hours. If gelatin capsules are used for embedding they should be dried at 60°C for an hour or so before use. Polythene capsules and embedding moulds or boats need not be pre-dried.

Method 1

Stock solution: 1:1 mixture of Araldite CY 212 and Hardener HY 964. This should be mixed thoroughly to ensure success, but do not trap air in the mixture as it may stiffen. A magnetic stirrer is ideal for the purpose. If the components are warmed to 60°, mixing will be facilitated. This stock solution is stable at room temperature for weeks. (If exceptionally hard tissue is to be cut a tenth of the HY 964 may be replaced by MNA)

Embedding medium:

20 ml stock solution
0·4 ml DY064 (catalyst or accelerator)
0·6 ml dibutyl phthalate (plasticizer)

This recipe yields a block of medium hardness. The consistency of the block is controlled by adding more or less of the plasticizing agent.

The bottle containing the embedding medium is attached to a turntable or placed on a roller-mixer and mixed at moderate speed for at least 2 hours before using. If there is medium left over after embedding the tissues, it may be kept for some weeks in a stoppered vial in the freezing compartment of a refrigerator or in a deep freeze as long as care is taken that moisture does not enter the vial. Just before use, it is warmed in a 60°C oven for 10 min to reduce viscosity, and to remove any moisture before opening.

Method 2

Follow the instructions for Method 1. The only changes involve the components of the stock solution and embedding medium.

Stock solution: 10 parts Araldite 502
7 parts HY964 (DDSA)

Embedding medium: Mix the stock solution with 1·5 to 2·0 per cent DY064 (DMP-30). If the DY064 is fresh 1·5 per cent suffices. As it ages 2·0 per cent may be needed. Mix thoroughly.

This recipe yields blocks of medium hardness. If harder or softer blocks are desired, alter the amount of hardener (HY964) in the stock solution.

Epon Embedding

Since Epon is somewhat hydroscopic, some difficulty has been encountered cutting Epon blocks, especially during damp weather. Stock solutions should be kept tightly covered to avoid uptake of water from the air. Infiltrating and embedding containers should be covered for the same reason.

Luft's (1961) modification of the technique is the one most generally used. Two stock solutions are needed, and will keep at 4°C for about 4 months:

1.	Epon 812	62 ml
	DDSA (HY964)	100 ml
2.	Epon 812	100 ml
	MNA	89 ml

Solution 1 yields very soft blocks, solution 2 very hard blocks, so they are mixed to obtain the desired degree of hardness. Seven parts mixture 1 to 3 parts mixture 2 yields a block of medium hardness.

After dehydration the tissue is placed for one hour in 1:1 mixture of ethyl alcohol and embedding mixture containing 1·5 to 2 per cent DY064 (DMP-30) as a catalyst. The tissue is then transferred to pre-dried gelatin capsules (or other embedding containers) containing fresh embedding medium plus catalyst and polymerized overnight at 60°C.

Left-over embedding medium can be stored in the freezing compartment of a refrigerator or a deep freeze as described for Araldite.

Epon–Araldite Mixtures

Mixtures of Epon and Araldite have been coming into increasing use. Epon penetrates more rapidly than Araldite; Araldite generally gives more reliable preservation of ultrastructure and has better cutting properties. Many workers feel that the 'best of both worlds' is attained by mixing the two. After dehydration and passage through epoxypropane, the tissue is infiltrated at room temperature.

Recipe 1 (Mollenhauer, 1964)

100 g Epon 812
55 g Araldite 502
180 g DDSA (HY964)
10 g BDMA (or DYO64) in the amount of 5 g/335g of resin mixture

After infiltration, embed in fresh medium at 60°C.

The epoxide equivalent of Araldite and Epon varies from one supplier to another, and it has been claimed that sectioning difficulties may be caused by this. Araldite 502 and Epon 812 supplied by Polysciences is recommended (by Polysciences!) for this particular recipe for that reason.

Recipe 2 (Anderson and André, 1968)

20 ml Araldite 502 (or CY 212)
25 ml Epon 812
60 ml DDSA (HY964)
1·5 to 2·5 per cent DMP-30 (DY064)

After infiltration, embed in fresh medium at 60°C.

VESTOPAL W EMBEDDING

This polyester embedding medium was introduced by Ryter and Kellenberger (1958) and is recommended for high resolution work by Sjöstrand (1969), although other high resolution electron microscopists do not agree that it is superior to the epoxy resins for this purpose (e.g. Porter, 1964). Vestopal W is particularly useful for embedding hard tissues.

Recipe for embedding mixture

Add to Vestopal W one per cent of tertiary butyl perbenzoate (initiator) and 0·5 per cent of cobalt naphthenate (activator). THE INITIATOR AND ACTIVATOR SHOULD NOT BE MIXED DIRECTLY TOGETHER OR THEY WILL EXPLODE.

All three ingredients should be protected from light and stored in a refrigerator. Vestopal W may be stored for several months, but the activator and initiator are said to deteriorate after about 2 months.

Since polyester resins are insoluble in ethanol, acetone is used as a dehydrating medium.

Dehydration: 30, 50, and 75 per cent acetone, 15 min each
90 per cent acetone 30 min
100 per cent acetone (dried over copper sulphate) 30 min

Infiltration: 3 : 1 dry acetone : Vestopal W, 30 min
1 : 1 dry acetone : Vestopal W, 30 min
1 : 3 dry acetone : Vestopal W, 30 min

Embedding: 12 to 24 hours in embedding mixture at 60°C

METHACRYLATE EMBEDDING

The methacrylates give off vapours which may be harmful if inhaled, and should be used in a fume cupboard.

Pure butyl methacrylate yields blocks which are too soft for satisfactory cutting, whereas methyl methacrylate yields blocks which are too hard. The two may be mixed together in varying proportions to give blocks of intermediate hardness according to the consistency of the tissue to be embedded.

The following recipe is usually successful with tissue of intermediate hardness:

9 parts butyl methacrylate
1 part methyl methacrylate
0·5 to 2·0 per cent benzoyl peroxide (catalyst)

Benzoyl peroxide decomposes explosively at 106°C, but is relatively safe at normal temperatures. Nevertheless in Britain it is frequently shipped as a moistened paste, and it is probably advisable to store it in the refrigerator. The small quantities to be used as a catalyst may be dried on filter paper at 50 to 60°C just before use.

After dehydration in ethanol (no passage through epoxy propane is necessary) the tissue is infiltrated by the following method:

1:1 mixture of ethyl alcohol and methacrylate without catalyst	30 min
Methacrylate mixture without catalyst	60 min with one change
Methacrylate mixture with catalyst	30 min
Prepolymerized methacrylate	overnight

(Prepolymerization is done by heating the embedding mixture plus catalyst in a conical flask at 90°C in a water bath with constant agitation until it becomes viscous.)

After infiltration, place the tissue in gelatin capsules (predried for 24 hours at 40 to 60°C before use) and cover with fresh prepolymerized methacrylate. Polymerization is then completed by exposing the capsules to ultraviolet light for 1 to 3 days in a cold room. The recommended wavelength for polymerization is 3650 Å (the same as used in ordinary sun lamps).

Water-Soluble Embedding Media

Only Leduc and Bernhard's (1967) modification of glycol methacrylate (GMA) embedding is given below. For references to other water soluble embedding media see references at the end of Chapter 8.

N.B.: Do not use osmium fixatives prior to GMA embedding, as osmium renders the tissue opaque to the ultraviolet light which is used for polymerization.

Fresh lots of GMA should be used, since spontaneous polymerization will occur with time. If a lot is not completely miscible with water, discard it. (Or use it to practise the art of prepolymerization, which is very tricky.)

Embedding mixture: 7 parts 97 per cent GMA in distilled water
3 parts butyl methacrylate containing 2 per cent Luperco* as a catalyst.

Dehydration and embedding are done in a cold room. Dehydration is accomplished in increasing concentrations of GMA instead of ethyl alcohol:

80 per cent GMA monomer in distilled water	20 min
97 per cent GMA monomer in distilled water	20 min
Unprepolymerized embedding mixture	20 min
Prepolymerized embedding mixture	overnight

Place in fresh prepolymerized embedding mixture in pre-dried gelatin capsules. (Do not use polythene capsules; they are opaque to ultraviolet light.) Leave capsules uncapped for 30 min to eliminate air bubbles. Then cap tightly so that a minimum of air contacts the embedding material. Polymerize with an ultraviolet lamp (3650 Å) for 1 to 3 days in a cold room. The tissue should be about 2·5 cm from the u.v. tube.

To bring about prepolymerization as recommended for GMA mixtures by Leduc and Bernhard (1967), heat a thin layer (1 cm deep) of embedding mixture in a conical flask over a bunsen flame with rapid swirling until strong convection currents or boiling is noted. Plunge flask immediately into an ice bath and agitate until mixture reaches temperature of the ice bath. The prepolymer is said to keep indefinitely in a deep freeze. Ours, we find, will not keep more than a few weeks.

WARNING: The reaction can easily get out of control. Keep the mouth of the flask pointed away from you, and expect the results of your first few attempts to be a useless mass of polymerized foam. The conical flask should of course, be made of Pyrex or other heat-resistant glass.

VI. STAINS FOR INCREASING THE CONTRAST OF ULTRATHIN SECTIONS

Only a few of the most commonly used stains are detailed below. For variants of these and for other stains see references in Chapters 4 and 8.

STAINS WHICH RESULT IN A GENERAL NON-SPECIFIC INCREASE IN CONTRAST

For a more thorough discussion of these staining techniques and their

* Luperco is not available in Britain, and Bernhard (personal communication 1971) says that it has no advantage over benzoyl peroxide which he now routinely uses. Moreover, Luperco is more hazardous to use than benzoyl peroxide.

variations, see Pease (1964, Chapter 7) and Glauert (1965).

Optimum staining time will vary with the embedding medium. The times quoted below have been found useful with Araldite embedding material, unless otherwise specified.

A convenient staining tray can be made by hollowing out three or four smooth depressions in a perspex strip the size of a microscope slide. Stain is dropped into each depression and grids can either be floated on the surface or immersed in the drop. If it is desirable to keep air from the stain, or to prevent evaporation during prolonged staining, a clean glass microscope slide can be carefully lowered at an angle onto the convex meniscus of the drops, thereby creating an air-tight seal. (Surface staining is, of course, not feasible if the drops are covered in this manner.)

An alternative method is to place the grid on or in a single drop of stain on the surface of dental wax in the bottom of a Petri dish. Moistened filter paper may be placed beneath the wax if a moist atmosphere is required. The dish is then covered and left for the required time. Depression slides may also be used for staining.

1. *Uranyl acetate*

A. *Aqueous:* 1 to 2 per cent aqueous solution. Stable for weeks at room temperature. Stain on a hotplate for 15 min if lead citrate is to follow. If not, stain 30 min to 1 hour on a hotplate with staining tray covered to prevent evaporation and consequent precipitation of stain upon the section. Wash grids *carefully* with a light jet of distilled water from a wash bottle to remove any precipitate. Touch non-section-bearing surface of grid to filter paper to remove water. A saturated aqueous solution of uranyl acetate may also be used, but should be filtrated or centrifuged before use. A saturated solution will, of course, precipitate if evaporation occurs.

B. *Alcoholic:* Saturated solution in 50 per cent alcohol; filter before use. Stain for 30 min on a hotplate, covered to prevent evaporation and consequent precipitation. The alcoholic solution does not keep well and should be freshly made. It is said to stain more rapidly than aqueous solutions. One to two per cent solutions are recommended by some workers.

2. *Lead Stains*

These stains tend to absorb carbon dioxide from the air and form precipitates of lead carbonate upon the sections. The stain should be carefully removed by a disposable Pasteur pipette from the centre of an undisturbed bottle. The pipette should be discarded after use. If surface staining is desired, the filter paper in the bottom of the Petri dish may be moistened with sodium hydroxide solution to remove CO_2

from the air. Otherwise, the grids may be immersed in staining tray depressions and an airtight seal formed by covering with a clean microscope slide. Lead stains of high pH (11 to 12) containing chelating agents are more stable and less likely to precipitate.

A. *Reynold's Lead Citrate Stain* (Reynolds, 1963). This is probably the most popular lead stain in current use. It gives a clear, delicate rendering of cellular structures and if reasonable care is taken precipitation is not a problem. The citrate ions present in the staining solution help to prevent precipitation by acting as chelating agents.

The reaction between lead citrate and sodium hydroxide is thought to result in a complex cation containing two lead atoms $(Pb(OH)_2Pb)^{2+}$. This cation, whose constituent atomic weights total 448·4, is thought to bind to reduced osmium, to hydroxyl groups in carbohydrate, and to ribonucleic acid.

Stain: 1·33 g lead nitrate
1·76 g sodium citrate $Na_3(C_6H_5O_2).2H_2O$
30 ml distilled water

Shake in a 50 ml volumetric flask and let stand 30 min with intermittent shaking to ensure complete conversion of lead nitrate to lead citrate. Then add 8 ml of 1 M NaOH and dilute to 50 ml with distilled water. Mix by inversion.

The pH is routinely found to be $12{\cdot}0 \pm 1$. The stain is stable up to 6 months and is said to 'mature' with age. Turbidity may be removed by centrifugation. *Staining time:* 5 to 30 min. The grids are rinsed before and after in stain with 0·02 M NaOH. After the final NaOH rinse they are rinsed twice in a gentle jet of distilled water from a wash bottle.

B. *Modification of Reynold's lead citrate stain* (Venable and Coggeshall (1965). Add 0·01–0·04 g lead citrate to 10 ml distilled water in a screw-topped centrifuge tube. Drop in 0·1 ml of 10 M NaOH. Close tube and shake vigorously until dissolved.

N.B. It is important to use only concentrated NaOH solutions or pellets in preparing the stain. Staining time varies from 10 seconds to 5 minutes. Double staining with aqueous uranyl acetate requires only a few seconds in each solution. Rinse stained sections by dipping rapidly twenty times in each of two or three vessels of distilled water. Air dry on filter paper.

Overstaining, too high a concentration of lead citrate, NaOH contaminated by atmospheric CO_2, or impure water, all can cause formation of large stain aggregates with consequent poor resolution.

C. *Karnovsky's Lead Stain, Method B* (Karnovsky, 1961). To 10 to 15 ml of 10 per cent sodium cacodylate in water lead monoxide (PbO) is added in excess, and the mixture is gently boiled for 15 min. After cooling, the mixture is filtered.

Two ml of the filtrate is placed in a 10 ml measuring cylinder and diluted to 10 ml with 10 per cent sodium cacodylate with stirring. NaOH (1 M) is added drop by drop with thorough stirring (for at least 15 seconds) between each drop. A faint cloudiness will form but will disappear with stirring after the final drop has been added. Do not add excess. Usually six drops is sufficient. When the solution is clear it is decanted into a 12 ml centrifuge tube, stoppered, and may be kept for several months. *Staining time:* Varies from 1 to 30 min. Rinse in distilled water.

D. *An Effective Double Staining Procedure for Routine Work*

1. Immerse or float grids on 2 per cent aqueous uranyl acetate for 15 min on a hotplate.
2. Gently wash grids with distilled water from a wash bottle.
3. Gently wash grids with 0·02 M NaOH from a wash bottle.
4. Stain with Reynold's lead citrate stain 30 min at room temperature.
5. Rinse gently with 0·02 M NaOH.
6. Rinse gently with distilled water.
7. Blot on filter paper; also blot forceps to remove any contamination by precipitate.
8. Repeat distilled water rinse.
9. Blot on filter paper.
10. Allow to dry in a dust-free place (e.g. in a petri dish) and store in gelatin capsules or special grid holders.

Staining Methods for Ultrathin Sections Which Claim Some Degree of Specificity

1. *Regressive Stain for RNA* (Monneron and Bernhard, 1969)

The tissue is fixed for 1 hour in phosphate-buffered aldehyde and embedded in Epon or GMA.

Staining routine for Epon sections:

1. 5 per cent aqueous uranyl acetate 1 min.
2. Rinse with distilled water.
3. 0·2 M ethylenediamine-tetraacetic acid (EDTA) 30 min.
4. Rinse with distilled water.
5. Stain with Reynold's lead citrate stain 1 min.

Staining routine for GMA sections:

The grids should be coated with Formvar or collodion before the sections are mounted.

1. 0·5 per cent aqueous uranyl acetate 1 min.
2. Rinse with distilled water.
3. 0·2 M EDTA in distilled water pH 7·0, 1 to 2 min only.
4. Rinse with distilled water.
5. Stain with Reynold's lead citrate stain for 1 min.

This is a 'regressive stain' in which chromatin is bleached by the chelating agent EDTA while RNA-containing structures remain stained. Results may be controlled by enzymic digestions as described by the originators of the technique.

2. *Phosphotungstic Acid (PTA) at Low pH for Polysaccharides and Glycoproteins* (Rambourg's 1969 modifications of Pease, 1966)
   Aldehyde fixation followed by GMA embedding is recommended.

   1. Float unmounted sections for 2 min on a solution of 1 per cent PTA in 10 per cent chromic acid (pH 0·3).
   2. Rinse briefly by floating sections on distilled water.
   3. Mount on Formvar or collodion coated grids.

Control: Acetylation for 12 hours at 37°C with acetic anhydride and pyridine (Marinozzi, 1968) abolishes the reaction. Material stained appears to be the same as that stained by the standard PAS reaction, except glycogen and lipid are extracted. Cell coat (glycocalyx), Golgi, lysosomes, basement membranes, brush orders and some secretion droplets are stained.

VII. STAINS FOR 1 TO 2 μm SECTIONS OF EPOXY RESIN- OR ACRYLIC-EMBEDDED MATERIALS

It must be remembered that staining reactions in material embedded in epoxy or acrylic plastics may be very different from those occuring in tissues from which paraffin embedding material has been removed. Specificity cannot be taken for granted.

Staining time and quality will vary with fixative and embedding material. Araldite generally takes longer to stain than the other plastics. The times given below have been found useful for Araldite sections, unless otherwise specified.

The same stain provided by different manufacturers may vary in effectiveness; different batches from the same manufacturer may also vary. To aid penetration of certain stains, Munger (1961) suggests soaking the sections in xylene for 1 hour, then passing them through decreasing concentrations of ethanol to water before staining.

Many stains will fade if exposed too long to light, or if dried on a warming tray. If permanent preparations are desired, it is usually best practice to air dry the stained sections at room temperature, mount them in D.P.X. or Permount, and allow the mounting medium to dry in the dark at room temperature. Mounting in epoxy glue may help to retard fading (Harrison, 1971).

1. *Quick methylene blue stain* (Philpott, 1966)

   1. Transfer sections to a drop of water or 10 per cent acetone on a microscope slide.

2. Adhere sections to slide by passing slide quickly through a bunsen flame. Sections should flatten nicely if water is not allowed to evaporate too quickly, and will adhere firmly to the glass.
3. Place on sections two drops of 1 per cent methylene blue in distilled water, and two drops of 1 per cent Borax (sodium tetraborate) in distilled water. The Borax raises the pH of the stain to approximately 9.
4. Pass slide through the flame 10 to 12 times. Do not allow stain to boil.
5. Rinse gently under a light stream of hot tap water. (If cold water is used, slide will break.)

The results of this method are rather variable; speed is its main recommendation. It is very useful for locating desired areas in the block of tissue and for general orientation. More uniform quality can be achieved by staining for 5 min or so on a hot plate, or immersing slides in a 1:1 solution of stain: Borax for $\frac{1}{2}$ to 2 hours, depending on the batch of stain used and the embedding medium.

2. *Other Dyes at High pH*

The basic dyes toluidine blue, azure II-methylene blue, crystal violet, safranin O or thionine may be substituted for methylene blue in recipe 1 (Richardson, Garrett and Finke, 1960).

In our laboratory we have found that a saturated solution of Giemsa or a 1 per cent solution of pyronine Y also stain well when mixed on the slide with 1 per cent Borax and flamed. If the pyronine Y is followed by 1 per cent toluidine blue or 1 per cent methylene blue mixed on the slide with 1 per cent Borax and flamed, a very pleasing dichrome stain is produced. The timing of the second stain is critical, and varies from one batch of stain to another.

3. *Methylene Blue–Basic Fuchsin Stain* (Aparicio and Marsden, 1968)

1. Stain with 1 per cent methylene blue in 1 per cent Borax on a hot plate at 70°C for 15 to 30 seconds.
2. Rinse in hot tap water and allow to cool.
3. Stain with 2 per cent basic fuchsin in distilled water for 1 to 2 min at room temperature.
4. Rinse in cold water.
5. Dry and mount in D.P.X.

(We found that strict adherence to this recipe gave unsatisfactory results with Araldite sections. However, if the sections are first soaked in xylene for an hour or more, passed through decreasing concentrations of ethanol to water, and the stains heated on the slide by flaming, the results are excellent.)

4. *A Periodic Acid–Schiff Technique*

A number of variants of the PAS technique have been reported successful with Epon and Araldite sections. A variant which we have found useful follows:

1. Dry sections to slide by flaming.
2. Soak in xylene at least 1 hour to facilitate penetration of stain.
3. Hydrate sections by passing them through decreasing concentrations of ethanol to water.
4. Oxidize for 15 min in 1 per cent periodic acid.
5. Rinse in distilled water.
6. Stain for 45 min in Schiff's reagent.
7. Rinse in two changes (2 min each) of freshly prepared bisulphite solution (5 ml of 10 per cent $K_2S_2O_5$, 5 ml M HCl, water to make 100 ml).
8. Rinse in two changes of distilled water.
9. Counterstain lightly with equal amounts of 1 per cent methylene blue and 1 per cent borax mixed on the slide and flamed. Four to five quick passes through the flame is usually sufficient. Rinse in hot water.
10. Dry in air at room temperature, protected from the light.
11. Mount in D.P.X.

VIII. PREPARATION OF FILMS FOR COATING GRIDS

The three most commonly used support films are made from collodion (nitrocellulose), Formvar (polyvinyl formal), or evaporated carbon. Evaporated carbon will form strong films as thin as 20 Å and is used mainly for high resolution work.

1. *Collodion Films*

If the collodion films prepared by the following methods have holes in them, the collodion solutions have probably become contaminated with water and should be replaced.

A. *Making a Film on a Water Surface* (After Drummond, 1950)

A circle of fine wire gauze (about 100 mesh/inch) is placed in the bottom of a shallow dish about 8 inches in diameter. The dish is filled with distilled water. A number of grids are placed on the gauze. The side on which one wants the film should be uppermost. Place two drops of 2 per cent collodion in amyl acetate from a pipette on the water at the centre of the dish. The solvent will evaporate and leave a film of collodion floating on the water. This first film will serve to clean the water surface of any dust particles, and is removed with a dissecting needle and discarded. *This is a very important step and should never be omitted.* A dirty support film can render a specimen useless.

A second film is formed in the same manner. The wire mesh is lifted out of the dish, picking up the film as it goes and thus covering the grids with the film. Alternatively the water may be drained off with a siphon with the same result. A large Petri dish with a drain tube and tap specially made for this purpose is available through some suppliers.

The grids are allowed to dry on the mesh in a dust-free place. They may then be picked up with tweezers. The film should tear at the edge of the grid, leaving the film on the grid intact. If difficulty is encountered, perforate the film around the grid with a needle or fine-pointed tweezers.

This method of preparing collodion films is valuable for preparing a large number of coated grids at one go. However, if the grids are not used soon after coating the films tend to become brittle and rupture.

B. *Floating the Film From a Slide*

This method tends to produce stronger films than Method A, and is the method of choice in many laboratories. It is useful for preparing a few coated grids at a time.

Prepare a 0·5 per cent solution of collodion in amyl acetate.

WARNING: Amyl acetate is volatile and may cause liver damage over a period of time. Keep the jar covered when not in use, or the solvent will evaporate and increase the percentage of collodion present. Stock solutions of a higher percentage (say 5 per cent) may be prepared and the 0·5 per cent working solution made from this by diluting with amyl acetate. The stock solution will remain stable for a month or so.

Put a small dot of Teepol or similar detergent on one face of a microscope slide and spread it thinly over the slide with a finger. Wipe off the excess with a clean cloth. (The slide should not be too clean or the film will not strip off.) Dip the slide into the collodion solution, remove, drain onto filter paper, and dry vertically in a dust-free place for about 10 min. With a razor blade or sharp-pointed tweezers score parallel to the edge of the slide on the face where Teepol is overlain by collodion (Fig. 17). Clean the surface of a dish of distilled water with a collodion film as described in Method A. Breathe on the slide. Dip it at an angle (about 30° from the horizontal) into the dish of distilled water. Do this slowly. The collodion film should float off onto the water.

The thickness of the film can be judged by the light reflecting from it. The thinnest films can be seen only by reflected light and will appear grey or silver-grey. Gold films are too thick. The percentage of collodion may be varied to get the proper thickness.

Grids are dropped onto the floating film. Light pressure on each grid with tweezers or a needle will ensure good contact between grid and film. A piece of lint-free Velin tissue is dropped onto the film. When this is picked up, the film and the grids come up with it. The Velin is placed, grid upward, in a covered petri dish to dry and can

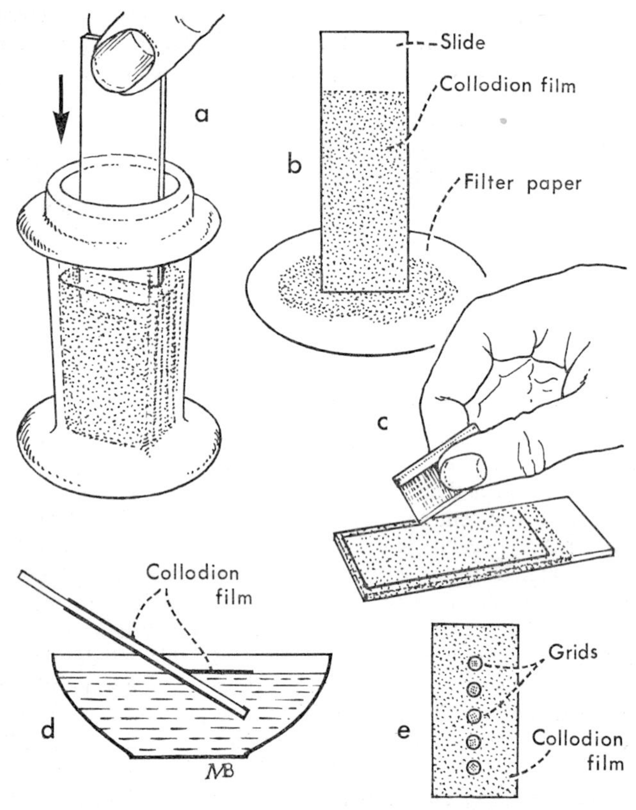

Fig. 17. Preparing a film on a glass slide. (*a*) Dip slide into collodion or Formvar solution. (*b*) Dry vertically. (*c*) Score parallel to edge of slide. (*d*) Float film off onto water surface. (*e*) Place grids on floating film.

remain there until the grids are needed. The grids may then be picked up with tweezers as described under Method A.

2. *Formvar Films*

Formvar films are stronger than collodion films, but may be more difficult to prepare and stock solutions must be renewed frequently. It is also important that the Formvar from which the stock solution is made is fairly fresh. If the film will not strip off the glass slide, check to see if the slides you are using are too clean. If this is not the cause of the trouble, try a fresh batch of Formvar.

Suitable Formvar films cannot be prepared by dropping onto a water surface. The films are always floated from a slide.

Prepare a solution of 0·3 to 1 per cent Formvar in ethylene dichloride *or* chloroform. Keep the jar covered when not in use, or the solvent will evaporate. Follow the directions for floating collodion films off slides (Method 1 B, above). Vary the percentage of the formvar until the required thickness is attained.

If chloroform is used as the solvent it is advisable to use a tall dipping jar so that the dipped slide may be drained before removing it from the jar. It is thus drained in an atmosphere nearly saturated with chloroform, which prevents formation of an uneven film through too rapid and uneven evaporation of the solvent (Pease, 1964, p. 198).

3. *Carbon Films*

Carbon films are stronger and more stable in the electron beam than either collodion or Formvar. They are also more suitable for high resolution work, it being possible to produce a carbon film as thin at 20 Å.

A. Either Formvar or collodion films on grids may be strengthened if necessary by deposition of a thin layer of carbon upon them by vacuum evaporation (for details of the carbon evaporation technique see Bradley, 1954, 1965).

B. A thin film of pure carbon for high resolution work may be obtained by depositing a layer of carbon on collodion or Formvar-coated grids and then dissolving away the plastic with the appropriate solvent (amyl acetate for collodion; chloroform for Formvar). Immerse the grids in the solvent for half an hour, then place in fresh solvent for another 10 min.

4. *Preparation of a Holey Carbon Film For Use in Performance-Testing the Microscope* (Adapted from Harris, 1962, by AEI Consultant Laboratory)

The addition of a small amount of water to a Formvar solution (in this case by the addition of a small amount of 50 per cent glycerol) will cause a film with holes to form.

1. Add 0·25 ml of 50 per cent glycerol to 30 ml of 0·5 per cent Formvar in chloroform.
2. Shake the solution by hand to disperse the glycerine in the Formvar. Just before use, agitate the solution in a sonicator for 5 min.
3. Prepare coated grids as described under Collodion Films Method B.
4. Allow the coated grids on the Velin tissue to dry for 30 min in a dust-free place.
5. Place on a filter wick in methanol for 15 min. (A filter wick is simply a stack of filter paper in a covered jar with its bottom immersed in methanol.) This dissolves away the glycerol and water.

6. Allow to dry.
7. Coat with a layer of carbon by vacuum evaporation. (The Velin tissue degasses quickly so the grids may be carboned without removal from the Velin.)
8. Dissolve away the Formvar by immersion of the grids in chloroform for ½ hour. Then place in fresh chloroform for 10 min. (This step is important, since a thick Formvar–carbon film will give rise to chromatic effects.)

IX. CLEANING METHODS

(Cleaning methods employed during routine maintenance of the electron microscope are described in Chapter 10.)

Cleaning Glassware

Teepol is a popular and effective detergent for cleaning glassware for routine use. It should be thoroughly rinsed off (three rinses of deionized water followed by three of distilled water). Teepol will not, however, remove all traces of the more stubbornly adhering organic materials, greases, or radioactive contaminants.

Acid Cleaning Solutions

Following detergent cleaning, badly contaminated glassware, or glassware such as bottles employed for fixing tissues, and items used in conjunction with enzyme histochemistry or autoradiography may be placed in an acid bath. The acid must, of course, be very carefully handled in order to avoid contact with the skin or splashing into the eyes. A commonly employed acid mixture is:

Potassium dichromate	100 g
Concentrated sulphuric acid	250 ml
Water	750 ml

Always add the acid slowly with stirring to the dichromate-water solution rather than vice versa. Otherwise the mixture may boil and spatter. Make it up in a heat-resistant flask in a sink.

The length of time that glassware should remain in the acid bath depends upon the nature of the contaminants on the glassware and on the concentration of the acid. Long soaking may be necessary, and some contaminants remain completely resistant to the corrosive action of acid. The glassware itself, however, may be attacked by the acid. This is particularly detrimental in the case of calibrated glassware.

If glassware is soaked in acid cleaning solution it must be thoroughly rinsed to remove all traces of acid. Some types of glass retain acid more easily than others. After a preliminary rinse in deionized water it is safest to leave the glassware in fresh distilled water for a time to allow any acid still adherent to the glass to soak off. Then rinse again in fresh glass-distilled water.

Decon 75 is an extremely efficient concentrated glass cleaning solution which has been found to be more effective than chromic acid in removing certain organic material, greases, radioactive contaminants, and even such tenacious contaminants as silicones, polyvinyl alcohol, polyvinyl acetate and polymethyl methacrylate (Bradshaw, 1970). It does not attack glassware and is completely removed by four rinses in deionized water. This should, in the case of glassware used for electron microscopy, be followed by a rinse in glass distilled water.

Decon 75 is non-toxic and is not hazardous to store or to handle, although gloves should be worn to avoid excessive drying of the skin. It is biodegradable, unlike many detergents, and thus presents no disposal problem. It is expensive, but is used in diluted form (2 per cent to 7 per cent solutions for most purposes; 20 per cent for radioactive contaminants). It should always be used above room temperature when gross contamination is being removed. The solutions are re-usable until the pH drops below 9·5. It would seem, in view of its effectiveness, that Decon 75 is an extravagance which is well worth it for glassware used for critical experiments.

Cleaning of Specimen Support Grids

Grids may be dirty when they arrive from the manufacturer, or may have become so in the laboratory. Dirty grids will contaminate the specimens which are picked up on them. Some forms of dirt will make picking up sections from the trough almost impossible.

Pease (1964, p. 215) recommends cleaning grids by shaking them up with a detergent solution in a 200 ml conical flask, rinsing well with several changes of distilled water followed by Analar acetone. The acetone is decanted and the grids dumped on large sheets of filter paper to dry in a dust-free place.

If expensive molybdenum, gold or other precious metal grids are to be re-used, the cleaning procedure will vary with what has been on the grid. Collodion or Formvar films may be removed with the proper solvent, e.g. amyl acetate or chloroform. Immerse in the solvent for one half hour, then in fresh solvent for 10 min. Rinse thoroughly with ethanol. If this does not result in complete removal of the film, sonication for 2 min in the solvent followed by 2 min in ethanol may be required. Spread the grids out on large pieces of filter paper or Velin tissue and dry in a dust-free place. Inspect each grid under the high dry objective of a light microscope to be sure that it is clean and fit for re-use.

X. PHOTOGRAPHY

The following photographic materials can be recommended for electron microscopic work:

Plates	Ilford	EM4 (fast), EM5 (average), EM6 (slow) Special Lantern Contrasty (fast)
	Kodak	Electron Image Plates (available from U.S.A.). Rather expensive to import
	Agfa-Gevaert	Scientia 23D50
Sheet film	Ilford	N7E50 6·5×9·0 cm
	Kodak	Electron Microscope Film (available from U.S.A. 20 boxes of 100)
	Agfa-Gevaert	Scientia 23D56
Roll film	Ilford	N4E50 (available from stock if on special list, by application)
	Agfa-Gevaert	Scientia 22D50 or 19D50
Developers	Ilford	PQ Universal (diluted 1:20)
	Kodak	DX 80 (diluted 1:10)
	May & Baker	Teknol (diluted 1:20)
All are liquid concentrates, convenient to use		
Fixatives	Ilford	Hypam
	Kodak	Rapid Fixer with Hardener
	May & Baker	Amfix and Super Amfix
Safelights		Ilford 902S or 904F, or Kodak OB

Recipe and Directions for Use of Farmer's Reducer

If a plate or film has not been too heavily overexposed, its density can be reduced to a useful level with Farmer's Reducer:

Solution 1	Sodium thiosulphate (hypo)	240 g
	Water to make	1000 ml
Solution 2	Potassium ferricyanide	19 g
	Water to make	250 ml

Store the solutions separately and mix immediately before use in the following proportions:

1 part Solution 2
4 parts Solution 1
32 parts water

The negative is immersed in the solution, then removed and checked every 15 seconds or so to see how reduction is progressing. When a satisfactory degree of reduction has been attained, rinse the negative in water, re-fix, rinse and dry.

XI. LIST OF SUPPLIERS OF MATERIALS USED IN ELECTRON MICROSCOPY

These suppliers will provide catalogues, information sheets and price lists upon request.

A. General Suppliers of Materials Used in Electron Microscopy. These suppliers provide a wide range of materials.

Great Britain	United States of America
Polaron Instruments. Ltd., 4 Shakespeare Road, Finchley, London N3	Polysciences, Inc. Paul Valley Industrial Park, Warrington, Pennsylvania 18976 (Polysciences is the American agent for all items produced by Polaron, London and will supply a separate Polaron catalogue on request. Polysciences is also an authorized Kodak agent)
Taab Laboratories, 52 Kidmore End Road, Emmer Green, Reading	Better Equipment for Electron Microscopy, Inc. (BEEM), P. O. Box 132, Jerome Ave. Station. Bronx, New York 10468
George T. Gurr, Ltd., Carlisle Road, The Hyde, London NW 9	Ladd Research Industries, Inc., P. O. Box 901, Area Code 802, Burlington, Vermont 05401
	Ernest F. Fullam, Inc., P. O. Box 444, Schenectady, New York
	Electron Microscopy Sciences, Box 251, Fort Washington, Pennysylvania 19034

B. Items listed below are those which are either not usually obtained through the general suppliers, or may be obtained in special grades or types, or more cheaply, elsewhere:

FIXATIVES

Osmium Tetroxide

Johnson & Matthey Chemicals, Ltd., 74 Hatton Garden, London EC1 (more cheaply than from general suppliers. Carbon rods and metals for evaporation also supplied)	Electron Microscopy Sciences, Box 251, Fort Washington, Pennsylvania 19034 (more cheaply than from most other suppliers)

Glutaraldehyde

Special 'EM Grade' glutaraldehyde is available from the general suppliers, usually as a 25 per cent solution. An especially pure grade is also available from Fisher Scientific (U.S.A.) through its U.K. agents, Eastman Kodak. It is designated G-151, 50 per cent, biological grade. The addresses:

Kodak, Limited, Chemical Division, Kirby, Liverpool	Fisher Scientific, 633 Greenwich Street, New York, New York 10014

EMBEDDING MATERIALS AND ACCESSORIES

Araldite

CIBA, Ltd., Duxford, Cambridge (also from general suppliers)	General suppliers

Epon 812

General suppliers	E. V. Roberts & Associates, Inc., 9601 West Jefferson Boulevard, Culver City, California 90230 (also from general suppliers)

Ultraviolet lamps for polymerization of methacrylates

P. W. Allen & Company, 253 Liverpool Road, London W1 (A405 is a useful model)	Ernest F. Fullam, Inc., P. O. Box 444, Schenectady, N. Y. (G.E. black light lamp No. F6T5, BL is suitable)

Gelatin capsules

Parke, Davis & Co., Ltd., Hounslow, Middlesex (or order through local chemist) No. 5 for grid storage. No. 0 for embedding	Eli Lilly & Co., Indianapolis, Indiana (or local druggist suppliers)

Micromoulds for embedding

Micron Associates, 6 Boston Gardens, Brentford, Middlesex	EMicron Research Ltd., Box 187, Postal Station 'T', Toronto 19, Ontario, Canada

MATERIALS FOR PREPARATION OF THIN SECTIONS

Diamond knives

Polaron Instruments, Ltd., 4 Shakespeare Road, Finchley, London N3 G–Fe–Ri De Guido Sasso & Co., Via Fontano Unica, Frosinone, Italy	E. I. Dupont de Nemours & Co., Inc., Instrument Products Division, 2102 North East Boulevard, Wilmington, Delaware 19898 (also from Ladd Research Industries and Ernest F. Fullam—addresses under general suppliers)

Specimen grids

Polaron Instruments, Ltd., 4 Shakespeare Road, Finchley, London N3 Mason and Morton, Ltd., Fir Tree House, Headstone Drive, Wealdstone, Harrow, Middlesex HA3 5QS Smethurst Highlight, Sidcot Heaton, Bolton, Lancashire LKB Instruments, Ltd., LKB House, 232 Addington Road, Selsdon, S. Croydon, Surrey	From general suppliers, but more cheaply from: Mason and Morton, Ltd., Fir Tree House, Headstone Drive, Wealdstone, Harrow, Middlesex HA3 5QS, England

Formvar

Shawinigan, Ltd. Marlow House, Lloyds Avenue, London EC3 (specify polyvinyl formal, Formvar 1595E)	Shawinigan Products Co., 350 Firth Avenue, New York, New York (specify Formvar E) Ernest F. Fullam, Inc., P. O. Box 444, Schenectady, New York 12301 (specify EM grade)

Ultramicrotomes and Accessories; Tissue Sectioners

Porter–Blum microtomes, Smith and Farquhar tissue sectioners, glassbreaking accessories

V. A. Howe & Co., Ltd., Sorvall Division, 88 Peterborough Road, London SW6	East Coast: Ivan Sorvall, Inc., Norwalk, Conn. 06852 West Coast: Ivan Sorvall, Inc., Sales Mart, 1485 Bayshore, San Francisco, Cal. 94124

Oxford vibratome

	Oxford Laboratories, 107 N. Bayshore Boulevard, San Mateo, California, 94401

Reichert ultramicrotomes

C. Reichert Company, 266 Bath Road, Slough, Buckinghamshire	American Optical Corporation, Scientific Instrument Division, Eggert and Sugar Roads, Buffalo, New York, 14215, U.S.A.

Cambridge Huxley ultramicrotomes and glassbreaking accessories

Cambridge Scientific Instruments, Ltd., Medical Unit, Histon Road, Cambridge CB4 3JA	LKB Instruments, Inc., 12221 Parklawn Drive, Rockville, Maryland 20852

LKB ultramicrotomes

LKB Instruments, Ltd., LKB House, 232 Addington Road, S. Croydon, Surrey, CR2 8YD	LKB Instruments, Inc., 12221 Parklawn Drive, Rockville, Maryland 20852

Cleaning Materials

Lint-free papers and cloths

Velin tissues General Paper and Box Mfg. Co. Ltd., Severn Road, Treforest Industrial Estate, Pontypridd, Glamorganshire	*Kim-wipes and Twill-jean cloths* Ladd Research Industries, Inc., P. O. Box 901, Area Code 802, Burlington, Vermont 05401

Detergents

Teepol Available from most chemical supply houses	Available from most laboratory supply houses
Decon concentrate Medical-Pharmaceutical Developments, Ltd., Ellen Street, Portslade, Brighton BN4 1EQ	

Cleaning solutions for use in ultrasonic units

The Pentone Co. Ltd., Cramlington, Northumberland (suppliers of Inhibisol)	L. & R. Manufacturing Co., 577 Elm Street, Kearny, N. J.
Quadralene Chemical Products, Ltd., Liversage Street, Derby DE1 2 LA (suppliers of Quadralene instrument cleaner)	Ernest F. Fullam, Inc., P. O. Box 444, Schenectady, New York 12301

PHOTOGRAPHIC SUPPLIES AND PRINT PROCESSING MACHINES

Kodak, Ltd., P.O. Box 114, Kodak House Kingsway, London WC2B 6TG	Polysciences, Inc., Paul Valley Industrial Park, Warrington, Pennsylvania 18976
Ilford, Ltd., Ilford Essex *or* (for print processing machines): Professional Products Division, Ilford, Ltd., 53–54 Berwick St., London W1	Ilford materials seldom used in U. S.
May and Baker, Dagenham, Essex RM10 7XS	Seldom used in U. S.
Agfa Gevaert, Great West Road, Brentford, Middlesex (supplies MD3 print processor)	Agfa Brovera supplied by photographic suppliers 3M Company, 3M Center, St. Paul, Minnesota 55101 (For print processing machines)

REFERENCES FOR APPENDIX

Anderson, W. A. & André, J. (1968). The extraction of some cell components with pronase and pepsin from thin sections of tissue embedded in an Epon-Araldite mixture, *Journal de Microscopie* **7**, 343–354.

Aparicio, S. R. & Marsden, P. (1968). A rapid methylene blue–basic fuchsin stain for semi-thin sections of peripheral nerve and other tissues. *Journal of Microscopy* **89**, 139–141.

Bennett, H. S. & Luft, J. H. (1959. *s*-Collidine as a base for buffering fixatives. *Journal of Biophysical and Biochemical Cytology*, **6**, 113–114.

Bradshaw, J. R. (1970). Comparison of surface-active agents. *Process Biochemistry* **5**, 19–20.

Caulfield, J. B. (1957). Effects of varying the vehicle for osmium tetroxide in tissue fixation. *Journal of Biophysical and Biochemical Cytology* **3**, 827.

Glauert, A. M. (1965). In *Techniques for Electron Microscopy* (Edited by D. H. Kay), Oxford: Blackwell Scientific Publications.

Drummond, D. G. (1950). The practice of electron microscopy: basic techniques. *Journal of the Royal Microscopical Society* **70**, 17–19.

Harris, W. J. (1162). Holey films for electron microscopy. *Nature, London* **196**, 499–500.

Harrison, G. A. (1971). Personal communication.

Hodgman, C. D., Weast, R. C. & Selby, S. M. (Eds.) (1970). *Handbook of Chemistry and Physics*. 5th edn., Cleveland: Chemical Rubber Publishing Co.

Karnovsky, M. J. (1961). Simple methods for 'staining with lead' at high pH in electron microscopy. *Journal of Biophysical and Biochemical Cytology* **11**, 729–732.

Karnovsky, J. J. (1965). A formaldehyde-glutaraldehyde fixative of high osmolarity for use in electron microscopy. *Journal of Cell Biology*, **27**, 137–138.

Leduc, E. H. & Bernhard, W. (1967). Recent modifications of the glycol methacrylate embedding procedure. *Journal of Ultrastructure Research* **19**, 196–199.

Luft, J. H. (1961). Improvements in epoxy resin embedding methods. *Journal of Biophysical and Biochemical Cytology* **9**, 409–414.

Marinozzi, V. (1968). In *Fourth European Regional Conference on Electron Microscopy*, Vol. **2**, pp. 55–56 (Edited by D. S. Bocciarelli), Rome: Tipographia Poliglotta Vaticana.

Millonig, G. (1962). In *Fifth International Congress on Electron Microscopy*, Vol. 2, p. 8 (Edited by S. S. Breese), New York: Academic Press.

Mollenhauer, H. H. (1964). Plastic embedding mixtures for use in electron microscopy. *Stain Technology* **39**, 111–114.

Monneron, A. & Bernhard, W. (1969). Fine structural organization of the interphase nucleus in some mammalian cells. *Journal of Ultrastructure Research* **27**, 266–288.

Munger, B. L. (1961). Staining methods applicable to sections of osmium-fixed tissue for light microscopy. *Journal of Biophysical and Biochemical Cytology* **11**, 502–506.

Palade, G. E. (1962). A study of fixation for electron microscopy. *Journal of Experimental Medicine* **95**, 285–298.

Pease, D. C. (1964). In *Histological Techniques for Electron Microscopy*, 2nd edn, pp. 52, 198, 215, New York and London: Academic Press.

Philpott, D. E. (1966). A rapid method for staining plastic-embedded tissue for light microscopy. *Scientific Instruments* **11**, 11–12.

Porter, K. R. (1964). In *Modern Developments in Electron Microscopy*, p. 139 (Edited by B. M. Siegel), New York and London: Academic Press.

Rambourg, A., Hernandez, W. & Leblond, C. P. (1969). Detection of complex carbohydrates in the Golgi apparatus of rat cells. *Journal of Cell Biology* **40**, 395–414.

Reynolds, E. S. (1963). The use of lead citrate at high pH as an electron-opaque stain in electron microscopy. *Journal of Cell Biology* **17**, 208–212.

Richardson, K. C., Jarrett, L. & Finke, E. H. (1960). Embedding in epoxy resins for ultrathin sectioning in electron microscopy. *Stain Technology* **35**, 313–323.

Ryter, A. & Kellenberger, E. (1958). L'inclusion au polyester pour l'ultramicrotomie. *Journal of Ultrastructure Research* **2**, 200–214.

Sabatini, D. D., Bensch, K. & Barrnett, R. J., (1963). Cytochemistry and electron microscopy—the preservation of cellular ultrastructure and enzymatic activity by aldehyde fixation. *Journal of Cell Biology* **17**, 19–58.

Sjöstrand, F. S. (1969). In *Physical Techniques in Biological Research*, 2nd edn, Vol. 3, Part C., p. 181 (Edited by A. W. Pollister), New York and London: Academic Press.

Venable, J. H. & Coggeshall, R. (1965). A simplified lead citrate stain for use in electron microscopy. *Journal of Cell Biology* **25**, 407–408.

Weibel, E. R. (1969). Stereological principles for morphometry in electron microscopic cytology. *International Review of Cytology* **26**, 262.

Index